Apparel and Textile Exports
Strategies for WTO Era

Apparel and Textile Exports

Strategies for WTO Era

Dr. Pradeep Joshi

B.Text., M.B.A., Ph.D. (IIT, Delhi)

Head (MDP/EDP) & Associate Professor,
National Institute of Fashion Technology, New Delhi

CBS

CBS PUBLISHERS & DISTRIBUTORS

NEW DELHI • BANGALORE

Soft Cover   ISBN : 81-239-1324-9
Hard Cover ISBN : 81-239-1325-7

First Edition : 2006

Reprint   :   2008

Publishing Director : Vinod K. Jain

Published by :
Satish Kumar Jain for CBS Publishers & Distributors,
4596/1-A, 11 Darya Ganj, New Delhi - 110 002 (India)
E-mail: cbspubs@vsnl.com • Website: www.cbspd.com

Branch Office :
2975, 17th Cross, K.R. Road, Bansankari 2nd Stage, Bangalore-70
Fax : 080-26771680 • E-mail : cbsbng@vsnl.net

Printed at :
Chaman Enterprises, New Delhi - 110 002

Dedicated

to

My Parents & Wife Seema

with Lots of Love

Foreword

The publication of the book, "Apparel & Textile Export: Strategies for WTO Era" is based on deep knowledge, sharpened experience, and expert career which Dr. Joshi has.

Basically, his book deals with the up-to-date research on topic of global competition in post-MFA era for the old and important apparel & textile industry of India. Opening up of the world market has created both challenges and opportunities. He indicated to Industry, Government, and Academia to prepare for such kind of coming namely 'Big Bang' after abolishing of MFA quotas due to WTO. After a certain period of incubation, the preparation time for touch with outside wind will be a critical phase to overcome such barriers and thus a valuable learning time to compete with them. In order to optimally utilize this opportunity, the strategy for long-term consistency and integration among players is necessary.

The book points out the problems of and Apparel and Textile Industry along with strategic options for being competitive in long run. For Apparel Industry, to shift into higher added value market, the author proposes the view of positioning at the higher in value chain, the increase of R&D investment, the technological development of finishing facility, the productivity improvement, the build-up of brand image, and the stronger connection with backward supply chain sections. For example, India's Apparel Industry is still concentrating "more on woven category where as the demand is increasing for knitted categories in last few years." Its consistent focus on low-end basic items market has been enforced by "non-availability of required fabric for high value items" in domestic supply market. Additionally also, for investment optimization, the need is for the economies of scale and the partnership among supply chain players.

For Textile Industry, the author focuses on non-availability of synthetic and blended fabric (compared with China, Taiwan, and South Korea), the technology problems at weaving and processing, and the unsuitable quality and non-optimized effort for export by large number of unorganized small sized fabric manufacturers. To compete with imported fabrics and to receive procurements from domestic apparel

Industry, they need the R&D for new material and production technologies. The productivity improvement "in terms of availability of wider width, consistent quality, required quality of finishing, consistency in lot/roll quantity, lot/roll quantity availability, count and construction availability", and quality improvement for global standard by investment, technology, new facility, and skill for industry.

Logical deployment of this book is matured due to his professional career. Tools such as SPSS, Excel and others have been used for statistical analysis. Factor analysis using PCA is also useful for objective explanation. Questionnaire design for data collection is very sophisticated.

However, quality is related with cost and lead-time. TQM (Total Quality Management) can be one of the steps for change to improve present situations. Perception or brand might partly depend on some objective measures such as ISO 9000 or JIS. For product development, the merchandising, product design, production technology, manufacturing, selling and after sales-service can become linked with each other by concurrent engineering or quick response system. Furthermore, Quality Functional Deployment (QFD) is also useful to design such system. Management techniques are more important than just capital investment, according to Japan's experience. Basically fashion is very difficult to forecast and seems has a sort of speculation. Then risk hedge technique as financial derivatives seems useful for valuable investment.

As India enters into the group of global market players as China or South Korea, she will become very important partner to Japan. While visiting Delhi or Bangalore, I always feel that this big and old country has great potential, as a hard working, enduring, vigorous, smart and tender youth. At World War II, Japan lost almost every thing. But in my area, Aichi Prefecture, main industry shifted from Textile into Automobile around 1974. Toyota also diversified from textile machines into automobile business in this region. Technology and know-how of textile machines were very useful to develop automobile. Then, when many low cost and good quality textiles and apparels came from South Korea or China, the work force relatively could be transferred into new business. But even so, Toyota has still kept the old business. New artificial fibers and textiles have been developed in Japan. The attitude for quality is always same, even if the situation is constantly changing.

Anyway, Japanese have experienced so many hurdles and are still learning many things after opening our country to the world about

130 years ago. In the ancient age, we learned Buddhism from India by way of China or Korea. I sincerely hope very long-living but still young India will achieve further success of economical development, in not only textile & apparel industries but also another industry with excellent contribution from young talent such as Dr. Joshi.

January 2006

DR. TAKAO FUJIWARA
Professor,
Division of Planning and Management
Toyohashi University of Technology, Japan

Preface

Indian apparel and textile sector is very critical to the Indian economy, being the highest net foreign exchange earner for the country. The phased removal of textile and apparel quotas since 1995 has catapulted Indian export firms in this sector into a new competitive environment. For more than 30 years, trade in this sector was governed by special regimes including STA, LTA and MFA, which provided for measures outside the normal GATT rules. The dismantling of the quota regime presents an opportunity as well as a threat. This impending reality brings the issue of competitiveness to the fore for all firms in the apparel and textile sector. It is imperative to understand the competitiveness of Indian apparel & textile firms in order to make an assessment of world trade in WTO era. I believe, quota elimination will enable retail buyers to focus on most competitive suppliers in terms of cost, quality and speed of delivery rather than being confined to those possessing quotas. The competitiveness of apparel & textile industry of a country will be determined by combination of various factors i.e. cost, supply chain, vertical integration, timely delivery, product adaptation & development as per requirements of target customers.

Though India has emerged as one of the main textile and clothing exporting countries, low and stagnant market share in the world textile and clothing trade, comparatively lower unit value realization, lack of presence in high value segment and aggressive performance of other Asian countries are concerns of Indian apparel & textile industry.

The approach in this book entails an effort to understand the Indian apparel & textile industry and environment scan, to identify the opportunities to be captured. Primary research with the structured survey of various stakeholders of industry i.e. apparel exporters, fabric manufacturers and buying houses has been conducted to identify steps & initiatives needs to be taken by industry and government to become competitive in WTO era. An attempt has been made to understand India's competitiveness in apparel & textile industry in WTO era with focus on assessment of competitiveness of fabric sector to fulfill the requirement of apparel industry.

This book has been organized in seven chapters. The first chapter presents an introduction to the evolution of apparel and textile industry and the changes in the global apparel & textile trade environment. Trade restrictions on apparel & textile industry have been discussed with focus on the implications of phasing out of the quotas along with brief trends in trade in WTO era. Second chapter is dedicated to the Indian apparel and textile industry, covering its composition & SWOT analysis. The third chapter presents a detailed account of competitiveness of Indian industry. The performance of various sectors of Indian apparel & textile industry is evaluated along with neighboring Asian countries so as to understand competitive position of India in world trade.

The fourth and the fifth chapters of the book elaborates on the findings of the survey of apparel manufacturers, buying houses and fabric manufacturers so as to understand competitiveness of Indian apparel & textile industry. The sixth chapter deals with opportunities in post MFA period along with expectations from trade and government to be competitive in world trade. The last chapter covers suggested strategies for Indian apparel & textile industry for WTO era.

The genesis of this book was in my research of world apparel and textile trade, which begins with my doctoral dissertation undertaken at I.I.T. Delhi and has continued since. The thought process initiated with my research in this area is published in various international journals and has benefitted exporters & working professionals of apparel and textile industry. The book, which marks an important place in an intellectual journey that I have been on for much of my professional life, grows out of my research, teaching at NIFT & industry interaction.

I would like to thank the Director General, NIFT, Ms Gauri Kumar, IAS, for motivation and encouragement in my research efforts. I am indebted to Dr. Sudhir K. Jain, Professor, Department of Management Studies and Dr. S.M. Ishtiaque, Professor, Department of Textile Technology, Indian Institute of Technology, Delhi for the guidance, support and words of advice during my doctoral research work.

I have received great support from apparel exporters, buying houses and fabric manufacturers representing apparel and textile industry in terms of timely and apt response during my long years of research in this endeavor. I would like to thank all of them. During the course of this research endeavor, I had the opportunity to interact and discuss various emerging issues in apparel & textile industry with different experts, academia, consultants and senior officers of various organizations. I would like to thank all of them for providing useful inputs.

I am grateful to professor Takao Fujiwara for writing forward of this book. I would like to thank him for sparing his valuable time & contributing in my endeavor.

I wish to thank Mr. H.S Poplai and the editorial, production and marketing team of CBS Publishers for taking great interest at every stage of publishing this book. I wish to thank all those who have directly or indirectly contributed in accomplishment of my book.

I have made an attempt to provide an understanding of Indian apparel & textile industry and suggested strategies for globalised trade environment in WTO era. This book identifies gaps in current strategy of apparel and textile industry, and focuses on unique elements offering competitive advantage to Indian industry. I hope this book will provide directions to Indian apparel & textile industry in devising strategies for WTO era.

PRADEEP JOSHI

Abbreviations

AEMA	Apparel Exporters and Manufacturers Association
AEPC	Apparel Export Promotion Council
ATC	Agreement on Textile and Clothing
CAGR	Compounded Annual Growth Rate
CBI	Caribbean Basin Initiative
CIRFS	International Rayon & Synthetic Fibers Committee
CIS	Commonwealth of Independent States
CITH	Textile & Clothing Information Center
CMP	Cut-Make-Pack
CMT	Cut-Make-Trim
DGCIS	Director General of Commercial Intelligence and Statistics
DGFT	Director General of Foreign Trade
DS	Department Stores
EC	European Commission
EEC	European Economic Community
EOU	Export Oriented Unit
EPC	Export Promotion Council
EPCG	Export Promotion Capital Goods
EPZ	Export Processing Zone
EU	European Union
EURATEX	European Apparel & Textile Organization
FCFS	First Come First Serve
FDI	Foreign Direct Investment
FOB	Free-on-Board
FYP	Five Year Plan
GATT	General Agreement on Tariffs and Trade
GL	Group Limit
IC	Industrialized Countries

ICRIER	Indian Council for Research on International Economic Relations
IIFT	Indian Institute of Foreign Trade
ILO	International Labour Organisation
IMF	International Monitory Fund
ISO	International Standards Organization
ITC	International Trade Centre
ITPO	India Trade Promotion Organization
IW	Importer-wholesaler
KSA	Kurt Salmon Associates
LDC	Least Developed Countries
LERMS	Liberalised Exchange Rate and Management Systems
LSC	Largest Supplier Country
LTA	Long term Agreement (regarding trade in cotton textiles)
MEP	Minimum Export Price
MFA	Multi-Fibre Arrangement
MFN	Most Favoured Nation
MOT	Ministry of Textiles
NAFTA	North American Free Trade Agreement
NIC	Newly Industrialized Countries
NIFT	National Institute of Fashion Technology
OBA	Outside Bilateral Agreement
OECD	Organization for Economic Co-operation and Development
ORC	Other Restrained Countries
Pc	Piece
PPE	Past Performance Entitlement
QR	Quantitative Restraints
RMG	Readymade Garments
SMEs	Small & Medium Enterprises
SSI	Small Scale Industries
STA	Short-term Agreement (regarding trade in cotton textiles)
SWOT	Strengths, Weaknesses, Opportunities and Threats
TEXPROCIL	Textile Export Promotion Council
TOI	Textile Outlook International
UVR	Unit Value Realization

Contents

Chapter I

Overview of Apparel & Textile Trade

The Indian apparel and textile sector is very critical to the Indian economy - as this sector accounts for around 8 percent of GDP, contributes 14 percent of the value addition in the manufacturing sector and more than 30 percent of the net export earnings of the country. It is the single-largest employer with an estimated workforce of 38 million. Industrial liberalization in the domestic economy since 1991 has been followed by changes in the global environment. Abolition of licensing controls on the Indian industry and the Uruguay round negotiations resulting in a ten year phase out of textile and clothing export quotas under the 'Agreement on Textiles and clothing (ATC)' posed a great opportunity and challenge to Indian industry. The roll down of textiles quotas was accompanied by roll down of custom duties and quantitative restrictions on import of wide range of textile and apparel products in to the Indian market. These sequential steps are leading to a convergence of the domestic and international markets for textiles and apparel. The domestic industry is no longer insulated from global competition. The phased removal of textile quotas since 1995 is catapulting Indian export firms into a new competitive environment. For more than 30 years, trade in this sector was governed by special regimes which provided for measures outside the normal GATT rules: the "Short Term arrangement" regarding international trade in cotton textiles (STA) in 1961, the "Long Term arrangement" regarding international trade in cotton textiles (LTA) from 1962 to 1973 and the arrangement regarding international trade in textiles, also known as the 'Multifibre Arrangement' or MFA, from 1974 to 1994. Through the transitional process embodied in the ATC, this sector is fully integrated into WTO rules from 1 January 2005. In particular, the extensive and complex system of bilateral restraints has come to an end and importing countries will no longer be able to discriminate between exporters by applying safeguard measures to individual countries. The industry will definitely face an acid test of survival post January 2005, with the opening of markets hitherto inaccessible and increased threats in the home market due to the availability of cheaper imports. With opening of textile and apparel markets in post-2004; the trade is set to offer opportunities as well as challenges to various players in world textile and apparel market. It can be termed as opportunity as with dismantling of quota system, restrictions shall be removed & more access to world consumption market is possible for manufacturers and exporters. The phasing out of quotas is threat to industry as assured market based upon past performance or international relationship etc. is ruled out.

Though India has emerged as one of the main textile and apparel exporting countries, low and stagnant market share in the world textile and apparel trade, comparatively lower unit value realization, lack of presence in high value segment and aggressive performance of China, Turkey, Bangladesh, Mexico, Korea and other Asian countries has made it necessary for Indian textile and apparel industry to understand the expectation and satisfaction levels of apparel exporters and buying houses representing buyers in destination market towards the availability and source of raw material i.e. fabric. The import of fabric particular synthetics or blends has increased significantly leading to affect Indian textiles (fabric) sector. India is sourcing raw material for apparel i.e. fabric and trims from countries i.e. china leading to becoming non-competitive in apparel trade. At the same time, due to lack of orders the fabric sector has also got affected which is affecting performance and focus of textile sector in India. Since raw material i.e. fabric consists of larger element in costing of apparel it is necessary to be competitive in fabric trade to remain competitive in world apparel trade. The perception of buying houses and apparel exporters regarding performance of fabric procured from various sources and reasons for performance are required to be understood. The absence of primary research in this area has been a retardant in the development of this sector.

The new trade regime with substantial reduction in tariff and duties, removal of a larger number items from the negative list of imports, liberal policies in favour of direct investment and considerably low duties for import of quota of capital goods and a host of such other steps have led the Indian economy into the main stream of global economy. More than ever before, attention is now on competitiveness based on marketing and manufacturing advantages rather than inherent factor advantages. Textile and apparel involve two of the most important products, which have been performing so far through comparative advantages. Apparel especially offers an opportunity for value added exports and therefore exploring the perceptions of importers and exporters to identify sources of competitive advantage become significant. The issue of competitiveness has been assuming far greater importance than ever as factor advantages are not proving to be of much help in a number of product categories. The differences are possible in total cost through efficient sourcing of indigenous materials at differential advantage, cost effective processes, productivity, range of services and intangibles like brand, quality and other non-price dimensions. Thus in a situation where quotas are no more controlling the supply mechanism, assessment of export capability and competitiveness achieve considerable importance. An in-depth research based on primary data collection through surveys of exporters and buying houses as representative of importers to identify the sources of competitive advantage becomes critical in this respect.

The India's textile and apparel industry has its inherent strengths in different stages of value chain alongwith a few weaknesses responsible for its present position in world trade. With liberalization of quotas, import of fabric has increased in last few years and is set to increase further due to weakness of India's textile sector in providing required fabric to apparel manufacturers. It is causing loss of opportunity and focus of textile and apparel industry has shifted to yarn manufacturing or manufacturing of greige fabric. This has affected competitiveness of fabric as well as apparel exporters and fabric manufacturers. The fabric as well as apparel manufactures have concentrated on woven sector while the world market for knitted items has grown up in last few years. Moreover, Indian exporters have targeted to basic segment rather than high value segment in fabric as well as apparel resulting in lesser earnings. The world textile and apparel trade in post-2004 shall be influenced by countries which are being able to cater to requirement of customers in time with quality product at right price. An attempt has been made in the book to identify the factors responsible for India's current position in world textile and apparel trade & measures to be taken to be competitive in world textile and apparel trade in post-quota period. In this book,the approach to study of competitiveness of textile & apparel industry in WTO era entails an effort to understand the situation of Indian textile & apparel industry and environment scan to identify the opportunities to be captured. A SWOT analysis provides gaps for textile and apparel industry. Primary research with the structured survey of various stakeholders of industry i.e. apparel exporters, fabric manufacturers and buying houses has been conducted to identify steps/ initiatives needed to fill the gap. The initiatives thus need to taken by industry and government to make Indian textile and apparel industry competitive in WTO era.

In the context of the competitive position of India, arrived at through empirical analysis of production in imports, growth rates and India's overall export performance and production this book purports to examine and identify the perceptions of apparel exporters and buying houses, fabric manufacturers on various elements related to export market of textile and apparel and competitiveness of textile (fabric) and apparel sector to reveal the sources of competitive advantage for Indian textile (fabric) and apparel exports. This research attempts through primary research of Indian apparel exporters and fabric manufacturers and buying houses to arrive at the perceptions and perceptual gaps in order to identify the sources of competitive advantage. An attempt has been made here to estimate India's competitiveness in textile and apparel industry in WTO era with focus on assessment of competitiveness of fabric sector to fulfill the requirement of apparel industry. These indicators should help us arrive at measures to increase India's competitiveness in textile and apparel industry.

In this study textile would be used to mean fabrics and shall also include fibres, yarn & made-ups at some places whereas apparel would stand for ready-made apparel. The terms garment, clothing and apparel would be used interchangeably.

1.1 WORLD TRADE IN APPAREL AND TEXTILE

The international trade in textile and apparel has a share of 5.1 percent in world merchandise trade in year 2004. The textile and apparel trade has registered a positive growth in last two decades. The growth rate of textile and apparel trade is higher than the growth rate of world agricultural trade (up 15 times in 40 years) but textile trade has grown slowly than the trade in total manufacturing goods (up 55 times in 40 years). The percentage share of textile and apparel in world trade has increased from 2.14 percent in year 1980 to 5.31 percent in year 1995 and to 5.7 percent in year 2000 (Table 1.1) showing a remarkable increase. The textile and apparel trade has been stagnant at 5.6 percentages in world trade for last few years, it has come to 5.1 percent of world trade in 2004.

Table 1.1 shows that the percentage share of textile in world merchandise trade has increased from 1.24 to 2.2 while percentage share of apparel in world trade has increased drastically from 0.9 percent to 2.9 percent in last two decades. The percentage change in world merchandise trade during year 1999-2004 has been of an increase of 99.20 percent while the share of textile and apparel has increased by 138.31 percent.

Table 1.1: Share of apparel and textile in world trade

Year	World merchandise trade (US$ bn)	percent Share of Textile and Apparel	percent Share of Textile	percent Share of Apparel
1980	4471.25	2.14	1.24	0.9
1990	5486.26	3.99	1.89	2.1
1995	5925.71	5.31	2.56	2.75
2000	6162.25	5.7	2.51	3.18
2001	6039.49	5.6	2.43	3.21
2002	6281.25	5.6	2.41	3.20
2003	7347.82	5.4	2.3	3.1
2004	8907.00	5.1	2.2	2.9
percent Change 2004/1980	99.20	138.31	78.86	222.82

Source: Compiled from WTO

The growth in share of textile trade is 78.86 percent while the apparel trade has increased by around 222.82 percent in corresponding period. It reflects more contribution of apparel in increasing share of textile and apparel in world trade.

Two decades ago, the value of world textile trade was US$ 54.99 bn (Table 1.2) while that of the world apparel trade was US$ 40.59 bn. As world textile trade has increased by two times and trade in apparel has grown more than four times, apparel trade has taken the lion's share of total world textile & apparel trade. Out of the US$ 453 billion of textile and apparel trade, slightly less than two third is trade in apparel while the rest is in textiles. The reason for increase in trade of apparel is preference of ready-to-wear rather than tailor made apparel by consumers leading to more demand of apparel in comparison to textile.

Table 1.2: **World trade in apparel and textiles US$ bn**

Year	Textile	Apparel	Textiles & Apparel
1980	54.99	40.59	95.58
1990	104.33	108.10	212.43
1991	108.86	117.06	225.92
1992	117.11	131.98	249.09
1993	112.31	128.55	240.86
1994	129.63	140.36	269.99
1995	151.58	158.30	309.88
1996	151.06	164.14	315.20
1997	157.73	182.28	340.01
1998	151.31	183.33	334.64
1999	146.23	184.60	330.83
2000	154.74	196.78	351.52
2001	146.98	195.03	342.01
2002	152	201	353
2003	169	226	359
2004	195	258	453
percent Change 2004/1980	254.60	535.62	373.94

Source: Compiled from WTO

In general, developing countries have a comparative advantage in textile and apparel trade. This advantage allows developing countries to diversify their exports beyond traditional primary commodities, whose production may be restrained by natural resources. As a leading labour-intensive manufacturing sector, the textile and apparel industry is often thought to represent the first base in economic growth and development in a country. Moreover, unlike the primary agricultural commodities that are often income inelastic, demand for textile and apparel commodities steadily grows in both developed and developing countries as countries become wealthier. This implies that for many developing countries there is room for future expansion of their production and export capacities.

In terms of the contribution to a country's economic development, many countries' experience shows that once export growth begin in the textile and apparel sectors, other steps in economic development follows. This transition has taken place in Korea, Taiwan, and now is happening in China, India and many other South and Southeast Asian, and Latin American countries. One reason is that there are strong linkages between the textile industry and other economic sectors both agricultural and non-agricultural. Growth in the textile sector benefits "upstream" agricultural or manufacturing sectors through increased demand for material inputs or machinery and equipment. In addition, the textile and apparel sectors depend on the presence of many modern economic activities. Through developing export-oriented textile and apparel industries, a country acquires other knowledge and skills such as marketing, advertising, transportation, and communication. These advances highlight the importance of the textile and apparel industries to a country's development process.

1.1.1 World Trade in Textile

(i) Import pattern

European Union is largest importer of textile followed by US, China and Mexico, Japan. The percentage share of European Union in world imports of textile has decreased from 46.5 percent in year 1980 to 33.0 percent in year 2004. The percentage share of US has increased from 4.5 percent in year 1980 to 10 percent in year 2004 (Table 1.3) showing a remarkable increase in import of textiles in US markets. The growth is more significant for China and Mexico where in last two decades the overall imports have increased drastically. The share of China in world import has increased from 1.9 percent to 7.4 percent of world import while the import of Mexico has increased from 0.2 to 2.8 percent of world trade in corresponding period.

The other countries including Mexico, Korea and Turkey have also become attractive destinations for textile exporters due to increase in their share in total world market.

Table 1.3: World's leading textile importers

	Value (US$ bn)	Share in world Imports (%)			
Importers	2004	1980	1990	2000	2004
European Union (25)	67.97	46.5	46.7	33.8	33.0
extra-EU (25) imports	20.99	14.0	13.2	9.9	10.2
United States	20.66	4.5	6.2	9.8	10.0
China	15.30	1.9	4.9	7.8	7.4
Hong Kong, China	14.11	–	–	–	–
retained imports	0.50	3.7	3.8	0.9	0.2
Mexico	5.79	0.2	0.9	3.6	2.8
Japan	5.60	2.9	3.8	3.0	2.7
Turkey	4.17	0.1	0.5	1.3	2.0
Canada	4.11	2.3	2.2	2.5	2.0
Korea, Republic of	3.38	0.7	1.8	2.1	1.6
Viet Nam	3.33	–	–	0.8	1.6
Romania	3.33	–	0.1	1.0	1.6
United Arab Emirates	2.15	0.8	0.9	1.3	1.2
Russian Federation	2.10	–	–	0.8	1.0
Australia	1.83	2.0	1.3	1.0	0.9
Thailand	1.81	0.3	0.8	1.0	0.9
Above 15	142.06	76.1	73.4	70.7	69.1

Source: WTO

(ii) Export pattern

European Union is largest exporters of textile followed by China, US, Korea, Taipei, Japan and India. The share of European Union in world exports has decreased from 49.4 percent to 36.6 percent during 1980-2004. The exports of China have increased many folds from 4.6 percent to 17.2 percent (Table 1.4) during the similar period. The other countries including Korea, Taipei, Pakistan, Turkey, and Indonesia have also registered a growth in textile exports. The percentage share of Korea in world textile trade has increased from 4 percent to 5.6 percent during 1980-2004 while the percentage share of Taipei has increased from 3.2 to 5.2 percent during the same period. The percentage share of India in world market is 4.0 percent of total textile exports, which has only marginally increased in last two decades.

The share of India in world textile exports was 2.4 percent in 1980, which declined to 2.1 percent in 1990 and subsequently increased to 4.0 percent in 2004.

Table 1.4: World's leading textile exporters

Exporters	*Value (US$ bn)* 2004	*Share in world exports (%)* 1980	1990	2000	2004
Eurpean Union (25)	71.29	49.4	48.7	36.5	36.6
extra-EU (25) exports	24.31	15.0	14.5	11.2	12.5
China	33.43	4.6	6.9	10.4	17.2
Hong Kong, China	14.30	–	–	–	–
domestic exports	0.68	1.7	2.1	0.8	0.4
re-exports	13.61	–	–	–	–
United States	11.99	6.8	4.8	7.1	6.2
Korea, Republic of	10.84	4.0	5.8	8.2	5.6
Taipei, Chinese	10.04	3.2	5.9	7.7	5.2
Japan	7.14	9.3	5.6	4.5	3.7
India	6.85	2.4	2.1	3.9	4.0
Turkey	6.43	0.6	1.4	2.4	3.3
Pakistan	6.12	1.6	2.6	2.9	3.1
Indonesia	3.15	0.1	1.2	2.3	1.6
Thailand	2.63	0.6	0.9	1.3	1.3
Canada	2.43	0.6	0.7	1.4	1.2
Mexico	2.24	0.2	0.7	1.7	1.1
Switzerland	1.60	2.8	2.5	1.0	0.8
Above 15	176.85	85.0	89.3	92.1	91.3

Source: WTO

1.1.2 World Trade in Apparel

(i) Import pattern

European Union is largest importer of apparel followed by US, Japan, Russia and Canada. The percentage share of US has increased from 16.4 percent in year 1980 to 24.0 percent in 1990 and further increased to 32.4 percent in 2000 and now is 28.0 percent in year 2004 (Table 1.5). The percentage share of European Union in world imports of apparel has reduced from 54.3 percent in 1980 to 50.6 percent in 1990 and further declined to 39.9 percent in 2000. The share of EU in world import is 45.0 percent in 2004. The other countries with high growth of import are Japan (3.6 to 8.0 percent during 1980-2004), Mexico (0.3 to 1.0 percent during 1980-2004) and Korea. The import pattern analysis indicates that US, EU, Japan contribute sizeable percentage of world apparel imports while all other countries have relatively smaller share in import of apparel.

Table 1.5: World's leading apparel importers

Importers	Value (US$ bn) 2004	Share in world Imports (%) 1980	1990	2000	2004
European Union (25)	121.66	54.3	50.6	39.9	45.0
extra-EU (25) imports	65.86	23.0	25.2	20.9	24.4
United States	75.73	16.4	24.0	32.4	28.0
Japan	21.69	3.6	7.8	9.5	8.0
Hong Kong, China	17.13	–	–	–	–
retained imports	0.17	0.9	0.7	0.8	0.1
Russian Federation	5.46	–	–	1.3	2.0
Canada	5.22	1.7	2.1	1.8	1.9
Switzerland	4.34	3.4	3.1	1.5	1.6
Korea, Republic of	2.75	0.0	0.1	0.6	1.0
Australia	2.67	0.8	0.6	0.9	1.0
Mexico	2.58	0.3	0.5	1.7	1.0
Singapore	2.06	0.3	0.8	0.9	0.8
retained imports	0.56	0.2	0.3	0.3	0.2
United Arab Emirates	2.05	0.6	0.5	0.7	0.8
Norway	1.67	1.7	1.1	0.6	0.6
China	1.54	0.1	0.0	0.6	0.6
Saudi Arabia	1.03	1.6	0.7	0.4	0.4
Above 15	250.61	85.8	92.8	93.7	93.0

Source: WTO

(ii) Export pattern

European Union is largest apparel exporter followed by China, Turkey, Mexico, US, India, Bangladesh and Indonesia. The percentage share of European Union has reduced from 42 percent in 1980 to 29.0 percent in 2004 (Table 1.6). While the percentage share of China has increased from 4 percent to 24.0 percent of total apparel exports. The other countries with high growth in exports of apparel are Turkey, Mexico, Bangladesh, Indonesia and Thailand. The percentage share of India has increased from 1.7 percent to 2.8 percent in world apparel exports between years 1980 to 2004.

Table 1.6: World's leading apparel exporters

Exporters	Value (US$ bn)	Share in world exports (%)			
	2004	1980	1990	2000	2004
European Union (25)	74.92	42.0	37.1	27.0	29.0
extra-EU (25) exports	19.13	10.4	10.5	6.9	7.4
China	61.86	4.0	8.9	18.3	24.0
Hong Kong, China	25.10	–	–	–	–
domestic exports	8.14	11.5	8.6	5.0	3.2
re-exports	16.96	–	–	–	–
Turkey	11.19	0.3	3.1	3.3	4.3
Mexico	7.20	0.0	0.5	4.4	2.8
India	6.62	1.7	2.3	3.1	2.8
United States	5.06	3.1	2.4	4.4	2.0
Romania	4.72	...	0.3	1.2	1.8
Indonesia	4.45	0.2	1.5	2.4	1.7
Bangladesh	4.44	0.0	0.6	2.0	1.7
Thailand	4.05	0.7	2.6	1.9	1.6
Viet Nam	3.98	...	...	0.9	1.5
Korea, Republic of	3.39	7.3	7.3	2.5	1.3
Tunisia	3.27	0.8	1.0	1.1	1.3
Pakistan	3.03	0.3	0.9	1.1	1.2
Above 15	206.32	71.3	77.5	78.6	80.3

Source: WTO

The above analysis indicates the growing importance of US as destination market for textile and apparel due to its large size and better realization in terms of prices. China has become significant player in world textile import as well as world apparel exports indicating emerging apparel industry, which is providing boost to their textile and apparel trade. Besides it, the countries i.e. Turkey, Mexico have registered a growth due to the emergence of trade blocks and indicate shifting pattern in world textile and apparel trade. The other countries i.e. India, Pakistan and Bangladesh are catering to low value segment but are having lesser share in the world market.

1.2 INDIA'S POSITION IN WORLD TRADE

The textiles and apparel is the largest manufacturing sector in India, accounting for around 14 percent of India's industrial output and a work force of 35 million persons. The sector also accounts for 27 percent of India's exports. World trade in textile has increased from US$ 104.33 (1990) to US$ 195 (2004) while India's textile exports has increased from US$ bn 2.18 to US$ bn 6.85 in year 2004 (Table 1.7). India's textile export has reached to US$ bn 5.89 in the year 2000. The percentage share of India in world textile trade has increased from 2.09 percent to 4 percent in world textile trade in year 2004.

Table 1.7: India's position in world textile trade

Year	World (value US$ bn)	India (value US$ bn)	%Share of India
1990	104.33	2.18	2.09
1995	151.58	4.35	2.88
1999	146.23	5.08	3.47
2000	154.74	5.89	3.82
2001	146.98	5.37	3.67
2002	152	5.38	3.53
2004	195	6.85	4.0
percent Change 2004/1990	86.90	214.22	91.38

Source: Compiled from WTO

Table 1.8 indicates that the percentage share of India in world apparel trade has marginally increased from 2.34 percent to 2.8 percent while in absolute terms it has increased more than 100 percent. In the year 2000 share of India in the world apparel has increased to 3.13 percent and in value terms to US$ bn 6.17.

The international trade in textile and apparel has a share of 5.1 percent in world merchandise trade in year 2004. The textile and apparel trade has registered a positive growth in last two decades. It also reflects more contribution of apparel in increasing share of textile and apparel in world trade. World apparel trade has increased from US$ bn 108.10 (year 1990) to US$ bn 258 (year 2004) showing a growth 138.66 percent while the growth of India's trade is from US$ bn 2.53 to US$ bn 6.62 during the same period (Table 1.8) i.e. 161.66 percent.

Table 1.8: India's position in world apparel trade

Year	World (value US$ bn)	India (value US$ bn)	%Share of India
1990	108.10	2.53	2.34
1995	158.30	4.11	2.5
1999	184.60	5.15	2.78
2000	196.78	6.17	3.13
2001	195.03	5.48	2.81
2002	201	5.6	2.78
2004	258	6.62	2.80
percent Change 2004/1990	138.66	161.66	19.65

Source: Compiled from WTO

It reflects that although India's apparel trade has increased in value terms but share in world market has not increased price. In other terms, India's competitive position in textile trade has improved while in apparel trade it is rather stagnant.

H.P. Bhattacharya (1992) classified developing countries and assessed the capabilities in the area of manufacturing and export of garments, which provide the broad indication of the global sourcing parameters. The classification has four groups based on certain characteristics:

Group I: The major East Asian suppliers consisting of South Korea, Taiwan, Hong Kong and China. Recent addition to this group is Turkey. These countries are capable of producing and exporting both finished textiles and garments of reasonably high quality (although predominantly in medium-price range) on their own accord.

Group II: Countries those are capable of exporting relatively low value-added and simple textiles such as grey yarn and unfinished cloth as well as low-fashion garments in low-to-medium price range. Among these countries are Brazil, Egypt, India, Pakistan, Indonesia etc.

Group III: Countries with industry to meet much of domestic demand for yarn and cloth but are not competitive in terms of price or quality in export market. Their garment export is dependent on imported fabrics and accessories. Notable in this group are East European countries, Malaysia, Philippines, Bangladesh, Sri Lanka, Tunisia, Morocco, many Latin American countries, Caribbean islands etc.

Group IV: Countries, which have some capability to meet part of domestic demand, but with limited possibility for export. Most African countries, Burma, Laos and Cambodia etc. fall in this group.

It can be seen from the above analysis that India is placed in group II among countries which are exporting low value added and simple garments in low to medium price range. India and Pakistan are examples of the countries whose garment exports overwhelmingly depend on cloth availability from decentralized (powerloom and handloom) sectors, which have limited capability of quality output and relatively low access to quality control and maintenance facilities. In order to move up the scale these countries need to strengthen their spinning, weaving and more importantly finishing facilities.

1.3 TRADE RESTRICTIONS ON APPAREL AND TEXTILE

Trade in apparel and textiles between developing and advanced countries has been subject to some form of restrictions since the mid-1930s. In the period immediately after World War II a major part of international trade was governed by complex national trade regimes. Post-war balance-of-payment difficulties in a number of developed countries were cited to justify high tariffs, complicated customs administrations, complex import licensing procedures and a wide range of quantitative restrictions. During the 1950s, however, trade restrictions were reduced in the wake of general liberalization efforts pursued in the GATT and the IMF.

The gradual removal of quantitative restrictions followed the easing of balance-of-payments difficulties in the developed countries. This coincided with the re-establishment of Japan in world trade in textiles and the emergence of a number of developing countries as exporters of textiles and to a lesser extent at that time, apparel. The developing countries, in particular, benefiting from access to raw materials and relatively low production costs, particularly wages began to rapidly increase the volume of exports of cotton textiles and apparel to the developed country markets. The sharp increase in low value imports of cotton textiles adversely affected investment and employment in the developed countries, which faced the prospect of rapid closure of production facilities in the sector leading to serious social problems. To alleviate the difficulties, some developed countries negotiated agreements with individual governments to limit the quantities of exports of cotton textiles or "voluntary export restraint" agreements, as they came to be known later. The United States imposed restrictions on textiles and apparel in 1935 through voluntary export restraints (VERs) to control shipments of cotton textiles from Japan. In the 1930s, some European countries including Britain and France imposed similar restraints on Japanese exports after World War II.

As far back as the 1950s, the textile industry in Western Europe and U.S.A. began to feel the pressure of intense competition of imports of cotton textiles from developing countries. It was a normal outcome of the process of economic development but instead of accepting the shifting pattern of comparative advantage and falling in line with the same, the affected countries decided to protect their domestic markets by abnormally high barriers. In May 1956, the Japanese Government was pressurized into applying voluntary restraints to 20 items of cotton textiles exported to the United States, followed by a five-year voluntary agreement, which resulted in Japan's share of the U.S. Export market falling from 55 percent in 1956 to 34 percent in 1961. In Western Europe, a "voluntary bilateral export restriction agreement" called the Noordwijk Agreement was reached in 1958 among the textile industries of the member countries of the newly established EEC; Austria, Norway and Switzerland, which pledged themselves not to re-export after processing to each other, gray cloth imported by them free of customs duty from Hong Kong, India, Pakistan, Japan and China. Later the Agreement was broadened to cover the exports of the socialist countries of Eastern Europe also. In the late 1950s, the textile industry of the United Kingdom entered into private bilateral agreements with industries in India, Hong Kong and Pakistan (known as the Lancashire pact) under which the industries in these countries agreed to limit their exports to the U.K. These private agreements with India and Hong Kong were later replaced by official agreements.

The competition form the developing countries were cited as the main cause of the problems faced by the textile industry in several developed countries although the real problem was the structural weakness of the industry. Even in 1960-61, when the pressures were building up for international action, a report by OECD observed that apart form some special branches such as shirt making, the competition of Asian and other developing countries had not been the main source of the difficulties of the European cotton industry. The textile industries in Europe as well as the United States however continued to exert considerable influence of their Governments for increased protective measures. On 2nd May 1961, the President of the United States of America while announcing a scheme of assistance for the domestic cotton textiles industry called for an international conference of all the principal textile exporting and importing countries to seek an international understanding to provide basis for trade that would avoid undue disruption of established industries.

The above initiative led to a meeting of the principal cotton textile importing and exporting countries under the auspices of GATT in

June 1961. Thus the ground was set for the introduction of discriminatory restraints, first, in the form of the "Short Term Arrangement" regarding international trade in cotton textiles in 1961 followed by the "Long Term Cotton Textiles Arrangement" (1962-73). The stated goals of these transitional arrangements were to significantly increase access to restricted markets to maintain orderly access to markets and to secure restraint in exporting countries to avoid disruption.

1.3.1 The Cotton Arrangements (1961-1973)

The LTA has prohibited participating countries from introducing new import restrictions or intensifying existing restrictions on cotton textiles, which would be inconsistent with the obligations under the GATT. Products subject to quota restrictions were stipulated to get expansion of access in the restricted markets. On the other hand the arrangement provided for safeguarding the importing countries against actual or threatened disruption of their domestic markets through one or more specified situations. The importing countries were also authorized to impose unilateral limits on imports if no agreement could be reached within 60 days of the receipt of a request by an exporting country for consultation on the level of exports. Such measures were, however, to be resorted to sparingly and limited to the precise products or groups or categories of products causing or threatening to cause market disruption. While LTA provided an assured access for the export of products of developing countries with a provision for expansion of such access, it was derogation form the basic principles of GATT which stood for protection of tariffs and for that matter a decreasing level of tariff.

1.3.2 The Multifibre Arrangement (1974 – 1994)

The MFA was first negotiated in 1974. It is an arrangement that allows eight industrialized country markets namely Australia, Canada, the European Union, Japan, Norway, Sweden, Switzerland and the United States to restrict imports and to restrain the exports of developing countries. Several other countries like Taiwan and Poland followed the MFA format but were not members. The movement from the LTA to the MFA-I during 1974-76 improved conditions for developing countries by providing higher export growth but the coverage of textile fibres in the new arrangement expanded from cotton to include wool, man-made fibres and rayon, as well as silk. Moreover, later versions of the MFA (MFA-II to MFA-IV) became more restrictive and even more complicated.

The main features of the agreement are:

(a) Integration of Textile and apparel products into the GATT/WTO rules and discipline thereby seeking to achieve non-discrimination and equal treatment for all the members' states.

(b) Procedures to address problems arising from circumvention like indirect shipping, re-routing and false declaration concerning countries and place of origin.

(c) Transition period for the phase out of MFA

(d) Improved market access than what was available under MFA regime.

(e) Establishment of Textile Monitoring Body (TMB) to supervise implementation of agreement and to deal with disputes till the existence of agreement on textile and apparel.

The MFA itself is a framework in which the developing countries agree to limit their exports to industrialized country markets. It operates under such principles as non-discrimination among exporting countries, minimum rates of export growth of 6 per cent in volume and flexibility provisions which include (i) swing provisions that permit the transferability of quotas among product categories, (ii) carry forward provisions that enable advance use of the following year's quotas and (ii) carry over provisions that allow transferability of unused quotas to the following year. The flexibility provisions are limited to 5 to 7 per cent of the quota in each category.

MFA was following the "Cotton Arrangements" through the provision of rules for imposition of restraints when sudden increase in imports was about to cause market disruption or threat thereof in importing countries. Extensions of this "temporary" measure were negotiated several times and new provisions were added and new products were also included. The restraints under the MFA were often negotiated or unilaterally imposed at relatively short intervals practically annually. Towards the end of the MFA, six participating countries Austria, Canada, EEC, Finland, Norway and United States were applying restraints under the MFA. The Arrangement was almost entirely used to restrict the imports from developing countries. Switzerland and Japan both members of the Arrangement never applied restraints towards the exporting countries under the MFA. Interestingly, Sweden became temporarily a non-quota country in August 1991 when all the quotas were abolished and the country left the MFA. This, however lasted only till 1.1.1995, as the country joined the EC and the EC quotas were imposed on the Swedish market. When the MFA came to an end on 31.12.1994; it had 44 members, less than half of the number of GATT members but the most significant producers/ traders in the international trade in textiles and apparel were part of it. China was not a contracting party of GATT, but it was a member of the MFA. Certain part of international trade in textiles and apparel was not restricted under the MFA, e.g. the trade among the developed countries. The MFA did not cover all the textile and apparel products and often the quota allocations were not fully utilized by exporting countries.

1.3.2.1 Developments under the MFA Regime

During its existence ,there were numerous changes and adaptations in the operation of the arrangement. Extensions of the MFA were negotiated a number of times, in the course of which new provisions were added and new products included. Furthermore, the growth rate of 6 per cent envisaged in the MFA was frequently sharply reduced in practice in bilateral agreements. The restraints under the MFA, which developed into a complex network of restrictions were negotiated or imposed in the case of unilateral actions, at short intervals, often every year or so. This period (1974-94) is characterized by:

(i) Patterns of shifting world production

A repeated pattern of expanding and shifting production occurred as investments in new facilities were made in the less-restricted or unrestricted exporters. This shift in production and export activity led to demands in the industrialized countries for yet more widespread restrictions in the new exporting countries which in turn brought about further shifts to unrestricted countries. While sustaining the performance of the originals predominant suppliers, this shifting production and export pattern stimulated the growth of industries in countries which, had it not been for the MFA restraint system, may not have entered the international market as exporters when they did. This also served to expand global production capacity.

(ii) Investment by developed countries

In reaction to ever-increasing exports by developing countries, developed countries made heavy investment in automated equipment, particularly in the textiles sector. As a result, the textile industries in the developed countries became capital-intensive one within the manufacturing sector. Extensive adjustment to improve productivity and efficiency in the face of intense international competition in both the textile and apparel industries of the developed countries led to a considerable decrease in the size of these industries in terms of both production facilities and employment.

(iii) Quality upgradation

Developing countries, subjected to quantitative restrictions, progressively upgraded the quality of their textile and apparel exports in order to maximize their economic gains. In shifting their exports to the higher end of the quality range, they forced competition to expand from the traditional area of lower value/high volume products to the more sophisticated ones where the developed countries were also concentrating their efforts.

(iv) Discrimination

The MFA and its cotton predecessor regimes were based on a selective approach, with discriminatory quantitative limitations on trade and were, therefore, a major departure from the rules of the "General Agreement". In fact, the GATT safeguard clause (Article XIX), while permitting quantitative measures in emergency cases to protect an industry from injury caused by "fairly traded" goods, does not permit discriminatory measures. As a result, not only was the development of textile and apparel exports from a number of countries limited, but also the exporting countries concerned had to forego their fundamental rights under the GATT with respect to non-discrimination, retaliation and compensation, in a sector of particular importance to them. For a large number of developing countries the expiration of the MFA on 31 December 1994, and the entry into force of the agreement on Textiles and apparel on the following day was most important result of the Uruguay Round.

(v) Circumvention of restraints

It has been observed that some exporters avoided restraints by the trans-shipment of goods through third countries and/ or false declarations of origin. This practice apparently took on extensive proportions in the later years of the MFA and led to increasing concern in the importing countries. The MFA and the bilateral agreements under its provisions created vested interests in both importing and exporting countries in the "guaranteed" market shared and the restraint rents associated with them. In these circumstances, the cost of protection was borne by the consumer in the form of higher prices.

1.3.3 Agreement on Textiles and Clothing (1995 – 2005)

The basic aim of the 'Agreement on Textile and Clothing' (ATC) is to secure the removal of restrictions applied by some developed countries to imports of textiles and apparel.

(i) Uruguay round agreement

The Uruguay round of multilateral trade negotiations was launched at Punta Del Est. in September 1986. One of the objectives of these negotiations has been to secure the eventual integration of the textiles and apparel sector into the GATT in a phased manner, on the basis of strengthened GATT rules and disciplines. The Uruguay round negotiations were concluded in December 1993 and, under the agreement on textiles and apparel, the developing country industry is to be integrated into the GATT 1994 over a transition period of ten years, beginning with the coming into force of the Agreement on 1 January 1995 along with the other agreements, including the agreement establishing the World Trade Organization (WTO).

(ii) World Trade Organization

The World Trade Organization (WTO) is the only International Body dealing with the rules of trade between the nations. At its heart are the WTO agreements, negotiated and signed by the bulk of the World's trading nations. These are essentially contracts binding Governments to keep their trade policies within agreed limits. Although negotiated and signed by Governments the aim is to help producers of goods and services, exporters and importers in doing their business. The main purpose is to help the trade flow as freely as possible so long as there are no undesirable side effects. It partly means removing obstacles. It also means ensuring the individuals, companies and Governments to know what the trade rules are around the world and giving them the assurance that there will be no sudden changes of policy, on the other hand the rules have to be 'transparent' and predictable. The agreements are drafted and signed by the community of trading nations and the WTO is to serve as a 'Forum for Trade negotiations' and to settle the disputes through some neutral procedure.

The World Trade Organization (WTO) was established on 1st January 1995 as a result of the 8th round of talks under GATT (1986-1994) known as Uruguay round. Till then GATT covered issues related to trade in goods. In the Uruguay Round, however, new agreements viz. 'General Agreement on Trade in Services' (GATS) and agreement on trade related aspects of intellectual property rights (TRIPS) were also negotiated. The WTO is the umbrella organization responsible for over seeing the implementation of all the multilateral and plurilateral agreements that have been negotiated in the Uruguay round and those that will be negotiated in the future. Its basic objectives are similar to those of GATT, which has been subsumed into WTO. These objectives have been expanded to give WTO a mandate to deal with trade in services. Furthermore, they clarify that in promoting economic development through the expansion of trade, adequate attention has to be paid in protecting and preserving the environment.

In brief the main objectives are:-

- Raising standards of living and incomes;
- Ensuring full employment;
- Expanding production and trade; and
- Allowing for the optimal use of world's resources.

The preamble extended these objectives:-

- To trade and services.
- To the need to promote 'sustainable development' and to protect and preserve the environment in a manner consistent with various levels of national economic development.
- To the need for positive efforts to ensure those developing countries, and especially the least developed among them, secure a better share of the growth in international trade.

The GATT 1994 refers to the original GATT, now referred to as the GATT 1947, plus the agreements, understandings and decisions related to trade in goods that were reached in the negotiations. Article 9 of the agreement on textiles and apparel states "This agreement and all restrictions there under shall stand terminated on the first day of the 121st month that the agreement establishing the WTO is in effect, on which date the textiles and apparel sector shall be fully integrated into the GATT 1994. There shall be no extension of this agreement".

The MFA permits the importing and exporting countries to enter into bilateral agreements to avoid sudden imposition of restrictions by the importing countries. In practice the important importing countries with all major exporting countries have generally entered into such agreements. Such agreements suit both the parties because they provide a measure of certainty to the trade. The exact quantities of various textile and apparel products which can be exported in a year are listed in these agreement, which as to provide after such matters as growth rates, carry over and carry forward of quotas to / from the next year, flexibility in swinging form one category to another and the procedural matters. India has at present bilateral agreements with US, Canada, European Community, Norway, Sweden, Austria and Finland.

According to the final text of the "Agreement on Textiles and clothing" each member country would on the commencement of the agreement (1st January, 1995) integrate into the GATT, products form the specific list annexed to the agreement which in 1990, accounted for not less than 16 per cent of the total volume of imports. At the beginning of phase two, on 1st January, 1988 products which accounted for not less than 17 per cent of 1990 imports would be integrated. On 1st January 2002 products, which accounted for a further 18 per cent of 1990 imports would be integrated. All remaining products would be integrated at the end of the transition period on 1st January 2005. In each of the first three stages, products shall be chosen from each of the four groups of (I) tops and yarns (ii) fabrics (iii) made ups and (iv) apparel.

The schedule for phasing out of Textile and apparel import quotas and integrating them to GATT rule over a period of 10 years is as follows (Table 1.9).

Table 1.9: Schedule for phase out of MFA

Phase	Percentage of products to be brought under GATT (including removal of any quotas)	How fast remaining quotas should open up, if 1994 rate was 6 per cent.
Phase 1 1 Jan 1995-31 Dec 1997	16 per cent (minimum, taking 1990 imports as base)	6.96 per cent year
Phase 2 1 Jan 1998- 31 Dec 2001	17 per cent	8.7 per cent per year
Phase 3 1 Jan 2002-31 Dec 2004	18 per cent	11.05 per cent per year
Phase 4 1 Jan 2005 Full integration into GATT (and final elimination of quotas)	49 per cent (maximum)	No Quotas Left

Source: WTO

The agreement provides for increasing the growth rates at each stage. During stage one (form the date of entry into force to the 36th months inclusive) the level of each restriction under MFA bilateral agreement in force for the 12 month period prior to the entry shall be increased annually by not less than the growth rate established for the respective category, increased by 16 per cent. During stage 2 (1998 to 2001 inclusive) annual growth rates during stage one shall be increased by 25 per cent and during stage three (2002 to 2004 inclusive) the growth rates during stage two grew by 27 per cent. Flexibility provision (swing, carry over, carry forwarded), as provided for in the MFA bilateral agreement for the 12 months period prior to the entry into GATT shall be brought into the GATT agreement as well.

Article 6 of the agreement contains a specific transitional safeguard mechanism, which can be applied to products not yet integrated into the GATT. Action under this article can be taken against an individual exporting country, if it is demonstrated by the importing country that overall imports of product were to enter that country in such increased quantities as to cause serious damage or actual threat there of to the domestic industry producing like and / or directly competitive products. It has been clarified that this serious damage or threat thereof must demonstrably be caused by increased quantities in total imports of that product and not by such other factors as technological changes or changes in consumer preferences. Furthermore, such restrictions will have to be applied to imports from all sources, and not on a discriminatory basis to imports from one or two countries as was the case with restrictions under MFA and is now under ATC.

Action under the safeguard mechanism can be taken either by mutual agreement, following consultations or unilaterally but subject to review by the 'Textiles Monitoring Body' (TMB). The level of restraints if fixed under this article, should be at a level not lower that the actual levels of exports or imports form the country concerned during the 12 months period ending two months before the month in which a request for consultation was made. Safeguard restraints can continue up to three years without extension or until the product is removed from the scope of the agreement, whichever comes first. The agreement provides for setting up a "Textile Monitoring Body" (TMB), to supervise the implementation of the agreement and to examine all measures taken under this provision and their conformity therewith. The TMB shall consist of a Chairman and 10 members, who shall be rotated at appropriate intervals, so as to be broadly representative of the member countries. The agreement also has provisions for special treatment to certain categories of countries such as those who have not been MFA members since 1986, new entrants, small suppliers, and least developed countries.

1.3.4 Integrating trade in textiles and clothing into WTO

When the WTO Agreement on Textiles and clothing, negotiated in the Uruguay Round, became operational on 1 January 1995, several importing Members [the United States, Canada and the European Union (15 countries) and Norway] had a total of 81 restraint agreements with WTO Members, comprising over a thousand individual quotas. In addition, there were 29 non-MFA agreements or unilateral measures imposing restrictions on imports of textiles.

From the strictly legal point of view, the maintenance of these restrictions was not consistent with GATT rules. However, MFA (negotiated within the framework of GATT) provided a legal cover for derogation from GATT discipline. The basic aim of the agreement on textiles and clothing, which carried over the quotas from MFA, is to integrate the trade in textiles and clothing into GATT rules and disciplines by requiring countries maintaining restrictions to eliminate them over a period of 10 years. After the expiry of the 10-year period, i.e. from 1 January 2005, it will not be possible for any member country to maintain restrictions on imports of textiles, unless it can justify them under the provisions of Article XIX of the GATT as interpreted by the WTO Agreement on Safeguards. Agreement on Textile and apparel is the only agreement under WTO which has its mortality, in-built in it through the process of integration. The genesis of the requirement for special treatment to textile and apparel under WTO dates back to the conflicting needs of industrialized-developed countries on one hand and developing countries on the other.

Since the GATT negotiations which started in 1947 achieved a token of success in removing tariff barriers in early 60s the industrialized developed countries sought to raise non-tariff barriers by way of quota restrictions and through multi-fibre arrangements (MFA), which allowed participants to employ special restraint mechanisms against injurious imports of specified goods from specified countries in certain defined circumstances, subject to their pursuing appropriate policies to encourage adjustment within their industries. The rights of importing countries to impose quota restrictions were balanced by their obligation to the low-wage suppliers to maintain annual growth rates and flexibility of the restraint levels. The MFA restraint regimes of the industrialized countries differed in coverage and protective effect due to fluctuating import patterns on account of relaxing and tightening their quotas and coverage. This led to vulnerability amongst low wage suppliers as also in the domestic producers in those countries due to the fact that overflow due to relaxation or lack of supplies due to tightening led to uncertainty in the market place, while the MFA was in effect.

Although, protection by the MFA provided apparel producers in the industrialized countries with time to adjust to increase import penetration, few of them made much progress in improving their competitiveness against low-cost goods. Nonetheless, in the Uruguay round of negotiations the cornerstone of the argument of these countries was that they needed protection provided by MFA for more years to protect employment levels and investments of their domestic textile industry. Developing countries on other hand had pitched their expectations on dismantling of quotas and increased access to the markets in these countries. The application of this agreement is not confined to MFA members only, and the same will apply to all the WTO members.

At each one of the phases of liberalizing the quotas; products to be integrated will encompass comprehensive groupings of Tops and Yarn, Fabrics and Made-up Textile Products and apparel. Members are otherwise free to select which products they want to liberalize at each stage of the process within the overall percentage. In this context members are unlikely to integrate products competing the most with domestic production as far as possible. Members who never applied MFA restrictions in the past may decide to integrate their products at once or to scale the integration to protect their rights to use the transitional safeguard measures. Further, the problems of circumvention as mentioned above is sought to be addressed by permitting plant visits and contacts, exchange of information and generally by finding a satisfactory solution through TMB. The schedule of integration of textile & apparel into GATT alongwith respective growth rates is shown in Table 1.10.

Table 1.10: Product integration schedule during MFA phase out

Category	Product integration schedule				Total
	Phase I 1.1.95	Phase II 1.1.98	Phase III 1.1.2002	Phase IV 1.1.2005	
Yarn	8.46%	8%	3.26%	2.64%	22.36%
Fabric	3.44%	2.51%	3.91%	12.19%	22.05%
Made-up	2.39%	4.54%	8.40%	2.55%	17.88%
apparel	1.92%	1.98%	2.54%	31.26%	37.71%
Total	16.21%	17.03%	18.11%	48.65%	100%

Source: WTO

The agreement provides for a safeguard clause, which is different from normal safeguard measures that can be taken under WTO, is the sense that these safeguard measures are available only during transition phase and that to for products, which are yet to be integrated. The maximum validity for such transition safeguard is only three years. The key product categories being imported from India alongwith respective quota growth rate in post WTO period for select countries indicate higher growth of quota in post – MFA scenario.

1.3.5 Implications of phase out of quotas

According to the GATT Secretariat the likely effects of the conclusion of the 'Uruguay Round' will by 2005 be some US$ 500 billion higher than before and based on 2002 trade figures (US$ 152 billion for textiles and US$ 205 billion for apparel). The additional potential growth for this sector could be as high as US$ 100 billion.

Implementation of ATC would lead to an increase in world trade with apparel being the primary category. It would greatly improve the supply situation, which would ultimately lead to a better price for imports. Developed countries in order to protect the interests of their local industry would be going in for bilateral agreements with their neighboring countries. Intra-regional sourcing by developed countries would dominate. In addition, they would increasingly use the tool of labour and environmental standards to retain their competitiveness in world trade. It is expected that more of antidumping duties and such non-tariff barriers would gain prominence as measures to restrict free trade. The ATC does not necessarily mean the end of quantitative restrictions after January 1, 2005; safeguard provision in the GATT itself permits resorting to quotas when domestic producers face serious injury from imports. The criteria for the assessment of serious injury are more stringent and the procedures for application more elaborate than the safeguard provision in the ATC.

Regional trade as against the traditional trading patterns is showing a major increase. Traditional apparel trade was primarily from the Asian subcontinent to the European Union and US. In the case of US, its imports from NAFTA, CBI countries as well as Latin American countries have shown a major increase. Likewise, the increase in the case of EU has been from the North African countries that have shown tremendous growth. The reasons that have led to the tremendous increase in regional trade include comparatively lower labour costs prevalent in neighboring countries of EU and US, a market for all intermediate textile products that EU & US make and better control and quick response possibilities.

The quota system has helped in the growth of Indian textile industry in so far as cheap Indian textile products found their way to the global market through reservations. The comparative advantage of cheap raw material and cheap labour, which is available in India, has helped it in emerging as a producer with distinction in handloom textiles, textiles of high quality standards as also in the textiles with low cost-low quality factors. India on its part is also committed to GATT to reduce import tariffs on seventeen textile products from the average rate of 85 percent in early 1990s to above 40 percent over a period of 10 years.

Thus, although markets have been opened to new exporting countries, the opportunities for export expansion have been limited and biased towards established exporting countries, such as the Republic of Korea, Taiwan (Taiwan, not a MFA member has bilateral agreements similar to the MFA with all importing countries) and Hong Kong. Among the newly industrializing economies (NIEs) in East Asia, Hong Kong and Korea accounted for around 40 percent of the available quotas in 1994, whereas countries in Southeast Asia accounted for only 14 per cent. Since bilateral agreements are usually based on historical performance large exporters have relatively large market shares and, therefore, it is difficult for new exporting countries with a low quota base to attain high market shares unless they have strong political bargaining power.

The apparel manufacturers have started building competence in specific product categories and markets. Hong Kong is working to supply branded products and be considered a hub for design, a move over from its manufacturing mould. Whereas Korea has moved out of appareling to specialize in delivering technical fabrics, Sri Lanka has created a niche and expertise in lingerie. Bangladesh meanwhile is manufacturing basic apparel, shirts trousers etc. and also concentrating on improving its fabric availability by providing warehousing facilities to its exporters. India on the other hand has developed strength in tops and trousers and is moving towards consolidating its factories for better management and productivity. Adding new machinery and training manpower seems to be the focus for improving quality and productivity to be able to offer right prices. The way business will be conducted beyond 2004 is shadowed by five major developments; consolidation of retailers worldwide, regional trade blocs, closing of manufacturing facilities in the US and the EU, and the China factor.

The agreement on textiles and clothing signed in December 1995 had a specific clause for OPT arrangements. US enter into OPT arrangements with Mexico, its NAFTA partner, and with the Caribbean Basin countries. EU, on the other hand, also entered into OPT arrangements but with Turkey and East European countries, chiefly Czechoslovakia and Bulgaria. The increase in OPT is observed due to –

a. Proximity which cuts down on time for dispatch of raw materials and receipt of finished apparel; and

b. Ability to exercise supervision on manufacturers.

In the last 5 to 8 years, the U.S. has increased apparel imports from Mexico under NAFTA and from Jamaica and Dominican Republic under the Caribbean Basin Initiative, while the EU has increased imports from the Central European countries. Most of this trade is free of quota restraints and hence is growing rapidly. This regional trade growth has built a constituency among a subset of exporting countries who now may find themselves more favorable to the MFA, since these countries now have export markets partly protected through MFA quotas from other suppliers. Second, Korea, Taiwan, and Hong Kong, which were the largest apparel exporters among developing countries 15-20 years ago, are now left with unused and unfilled MFA quotas for their exports, while countries like China, India, and Pakistan are facing tight quota restraints due to the fast growth in their exports.

The implications of the removal of quotas for various stakeholders are as follows:

i. Implications for different countries –

Within a more open market the relative competitiveness of countries depends mainly on:-

    a.   Labour costs
    b.   Supply of fabric, yarn and other raw materials
    c.   Infrastructure for transport and marketing
    d.   Nearness to markets

Importing countries will benefit from lower import prices due to abolishing of quota rent. They can source from the most efficient exporters. Resource allocation in importing countries will be improved from import competition. The quota rent will disappear, as will rent – seeking activities. Production and export will be rationalized and move to more efficient sectors.

Exporting countries may gain in one aspect and lose in the other depending mainly on their competitive advantage. The quota restricted competition and allowed less competitive exporters to exports more than their competitive shares. These less competitive exporters will lose their market shares. Countries whose exports have been limited by the MFA will gain from greater market access. However, exporting countries will face lower price after abolishing of the MFA.The outcome for any individual country will depend heavily on its policy response. Countries that take the opportunity to streamline their policies, and improve their competitiveness, are likely to increase their gains from quota abolition. To achieve competitiveness and improve supply response, the developing countries themselves have to endeavor to reduce business costs by improving governance, infrastructure & institutional supports, ensuring transparency and accountability and establishing the rule of law.

ii. Implications for companies –

Industrialized countries devised the MFA to protect their own apparel producers. At that time, most apparel was produced by local manufacturers and sold to local or national retails. Now the industry is mainly controlled by US and European based multinational companies, which own no production facilities themselves but manage an international network of suppliers. This includes big retailers such as Wal-Mart and brand based companies like Nike and Adidas. These companies, with the phase out of MFA, will benefit from an increasingly open market. As developing countries compete to become major suppliers, they can source more freely from the most profitable locations. Governments of developing countries are likely to invest public money in attracting overseas investment whilst the main profits will go to these overseas companies. Small companies now have to compete even more with international suppliers and many are facing threat of closing down. Larger manufactures are subcontracting an increasingly percentage of their production to lower wage economies.

iii. Implications for workers –

The removal of quotas will mean changes in the location of the industry. A less controlled system will mean that both countries and companies will be in a more direct competition to supply the world market. This will have implications for workers everywhere. There will be major shifts in the location of apparel production over the next decade. In the end, there could be greater stability as the location of apparel production becomes determined more by market forces than the arbitrary imposition of quotas. However, the initial impact may be highly disruptive to employment. The increase in competition at a global, national and local level is resulting in downward pressure on working conditions. With no quota restrictions, labour costs will be an even more significant factor. Developing country leaders rightly claim that the MFA has unfairly restricted exports vital to their development. However, development cannot be measured in only in terms of increased export earnings. Since the sector predominates in low-value production in the region where labour cost advantage provides the competitive edge, the rolling back of the assured quota could lead to the race to the bottom by putting pressure on producers to reduce labour cost to stay competitive.

1.3.6 Literature Survey on Trade Restrictions on Textile and Clothing

The emergence of Asian countries in textile and clothing trade is discussed by K.F.A.U. & N.Y. Chain (1997). According to him, before

World War II, advanced industrial countries in Western Europe and the US dominated the world economy and controlled most of the industrial production. The less-developed countries tended to concentrate in the production and supply of raw materials. Starting form the late 1940s, major textile and clothing industrial production has shifted out from developed countries and moved to Japan. Since then, Japan was the leader in industrialization and economic development in the Asian region. In 1970s, the high cost of production, labor shortages had compelled Japanese textile and clothing firms to invest their production in other Asian nations. Following Japan, Hong Kong, South Korea and Taiwan became three of the four Asian newly industrializing countries (NICs) with textile and clothing as their major export industry. In 1975 the average wage for US clothing workers was US$ 3.79 per hour, compared to US$ 0.75 in Hong Kong, US$ 0.29 in Taiwan and US$ 0.29 in Taiwan and US$ 0.22 in South Korea. NICs' cheap labor cost was the main attraction for the US and Japanese textile and clothing firms to locate their productions overseas. Moreover, the Asian NICs quickly added other benefits including improved quality level, flexibility of production and stylish merchandise. Thus the NICs can offer good quality textile and clothing products at a relatively lower price.

In response to the imposition of quotas by the US and other Western nation's, the traditional Asian textile and clothing producers have shifted their production to other less developed countries (LDCs) in Asia including China, Indonesia, Thailand, Pakistan, Sri Lanka and Vietnam since the 1980s. Boosted by Japan's foreign direct investment, China's clothing supplies accelerated to 74.7 percent of Japan's total clothing imports in 2000, compared with 27.4 percent in 1990. Hong Kong textile and clothing firms, in search for more quota holdings, also moved their production sites to mainland China. A 'Hong Kong Trade Development Council' survey conducted in 1998 explored that about 54 percent of Hong Kong's clothing exports were produced in mainland China. China has now become the largest textile and clothing exporter with other Asian countries such as India, Indonesia, Pakistan and Thailand also become the major exporters in the global textile and clothing market. The shifts of textile and clothing production to the Asian countries have witnessed the reliance of Asian economies on the textile and clothing industry to gain their early economic successes. Today, quite a number of the major global textile and clothing exporters are located in the Asian region.

The Uruguay round negotiations begun in 1986, in Punta del Este and ended with an agreement signed, in 1994 in Marrakech. The textile and clothing international trade changed significantly with the coming into force of this new agreement. Following a GATT Secretariat evaluation, the world trade liberalisation would allow an increase of 12 percent in manufacturing trade, in the year 2005, against the average of 4.1 percent during 1980-91. Textile and clothing items were considered likely to be the most benefited products with foreseen increasing rates of growth of 60 percent and 34%, respectively. Munir Ahmed (1997) studied the Pre and Post-Uruguay round talks on tariff reductions for textiles and clothing and its impact on the developing countries. The study dealt with the sensitivity of clothing to tariffs, in the major importing countries and the failure of Uruguay rounds negotiations on lowering of tariffs in these countries. It mostly dealt with the period of Pre and Post-Uruguay round of talks and concentrated on an empirical analysis of the impact of the MFA on lowering of tariffs, which would finally affect the UVR of most of the casual clothing product categories. The conclusion offered by the study was that while tariffs for all industrial products would be reduced by about 40 percent of the import value, as a result of the Uruguay round commitments by developed countries, tariffs on textiles and clothing might be reduced by only 2 percent. The study was seen from the perspective of the developing nations and drew a conclusion, that there would be tariff reductions, once the Multi-Fiber-Arrangement (MFA) would be lifted. It validated some of the findings made in the early 1990s, that there would be tariff reductions, and also drew the inference that the developing nations might not benefit really from the lifting of the quotas. The scope of the study was limited to only MFA liberalisation and did not offer a product category specialisation perspective.

The world market has been large and due to the World Trade Organisation (WTO), trade barriers have been falling, opening up tremendous opportunities and by the year 2005, quotas might have been removed from all categories of apparel. For success in global markets, Indian companies need to decide in which part of the value chain they should compete and what business system they will need to compete successfully, i.e., where and how to compete. A paper presented by McKinsey (1997) offered an optimistic picture for the Indian textile and apparel industry. "The world market abounds with attractive opportunities for Indian textiles companies. Indian companies are well placed to capture these opportunities. But they will first have to build a distinctive position for themselves. They have not done this, even in the Indian market", emphasised the study. The study dealt with the intervening period of lifting of the quotas, as well as with the benefits of liberalisation. It also featured the competitiveness of the

Indian apparel firms in terms of abundance of cheap labour, easy availability of cotton material, and high capital utilisation. It also provided
with a comparative analysis of China and India in casual wear exports.
As far as Indian apparel firms are concerned, it offered the prescription
of exporting, as a strategic imperative, citing the historical perspective
that the industry had been plagued by low profitability since 1976 and
operating margins have been low to negative and many textile mills
had closed down since then. The study also brought to the fore the
reference of China which was placed with India in the same bracket in
1980, but since then had made rapid strides, while India's share of
world textile trade continued to remain low. By 1994, China's share in
world's clothing exports had risen to 13 percent but India's share was
still 3%. An attempt had been made to study the impact of liberalisation of quotas by FICCI (2000). The study dealt with the future of the
world trade and suggested that the implementation of the agreement
on textiles and clothing (ATC) might lead to an increase in the world
trade, with apparel being the primary category. The conclusion was
that the importing countries would resort to bilateral agreements with
neighboring nations to protect the interests of their respective domestic industries. A significant level of 'Country of Origin' shift to low
cost countries would take place for the purpose of garment sourcing.
The study offered macro solutions to the export community as a whole
and did not offer solutions to textile and clothing manufacturers and
exporters to increase competitiveness in post MFA period.

1.4 TRENDS IN WTO ERA

The 'Multi-Fiber Arrangement' (MFA) has governed international trade
in textiles and clothing since 1974. The MFA enabled developed nations,
mainly the US, European Union and Canada to restrict imports from
developing countries, through a system of quotas. The 'Agreement on
Textiles and Clothing' (ATC) to abolish MFA quotas marked a
significant turnaround in the global textile trade. The ATC mandated
progressive phase out of import quotas established under MFA, and
the integration of textiles and clothing into the multilateral trading
system before January 2005. The trends in WTO era (Post Jan. 2005)
are discussed in this section.

1.4.1 US Market

Imports of textiles in US market (Table 1.11) has increased by 9.16
percent during Jan-July 2005 over corresponding period in year 2004.
Imports (value terms) have seen highest growth in China (23.55

Table 1.11: Import of Textiles in US market (Value in US$ mn)

Imports	(Jan-July) 2004	(Jan-July) 2005	percent change
World	10,766	11,752	9.16
Exporters:			
China P	3,219	3,977	23.55
Canada	930	925	-0.52
Pakistan	825	919	11.42
India	803	927	15.37
Mexico	648	668	3.21

Source: Compiled from Office of Textiles and Apparel, U.S. Department of Commerce

percent) followed by India (15.37 percent), Turkey (12.53 percent), Pakistan (11.42 percent) in this period. India has IInd position in import of textile preceded by China & closely followed by Canada, Pakistan. The imports from Korea, Taiwan, Thailand & Canada have decreased in the same period.

Import of apparel in US market (Table 1.12) has increased by 9.48 percent during Jan-July 2005 over corresponding period in year 2004. There is tremendous growth in imports from China (94.33 percent) followed by India (34.44 percent), Bangaladesh (21.54 percent) in this period. There is negative growth in imports from Mexico, Dominian Repubic etc. China, Mexico followed by India, Indonesia are key countries of imports for US apparel trade.

Table 1.12: Import of Apparel in US (Value in US$ mn)

Imports	(Jan-July) 2004	(Jan-July) 2005	percent change
World	35,466	38,827	9.48
Exporters:			
China P	4,673	9,081	94.33
Mexico	3,879	3,633	-6.34
India	1,324	1,780	34.44
Honduras	1,516	1,530	0.89
Vietnam	1,454	1,440	-0.95

Source: Compiled from Office of Textiles and Apparel, U.S. Department of Commerce

The analysis of apparel imports into US market in Jan-July 2005 indicates increase of 14.5 percent (volume) and increase of 11.31 percent (value) from corresponding period in 2004. The growth of cotton and wool apparel imports in value is similar while that of silk and other vegetable fiber has reduced. The US imports of Jan-July 2005 are US$ 32358.68 million (value) and 10451 mn.sq.mtrs in quantity. Cotton (US$ 20626.88 million) followed by MMF (US$ 9008.23 million) are leading categories of imports.

Table 1.13 indicates competitive position of India in apparel imports of US. china is highest gainer of phasing out of quotas in US market based upon data of Jan-July 2005 over 2004 with 125.19 percent growth in volume and 97.15 percent (value). The imports from India have increased by 32.31 percent (quantity) and 37.59 percent (value). The other gainers of WTO era are Bangladesh (22.73 percent value growth) followed by Sri Lanka (17.33 percent value growth) and Pakistan (11 percent value growth).

Table 1.13: India's competitive position

From	Jan-July 04		Jan-July 05		percent change	
	Qty (mn.sq.mtr)	Value (US$ mn)	Qty	Value	Qty	Value
India	312.70	1117.446	413.75	1537.52	32.31	37.59
China	1229.69	3755.49	2769.14	7403.97	125.19	97.15
Bangladesh	425.08	858.82	512.90	1054.03	20.66	22.73
Pakistan	237.43	497.00	265.18	551.66	11.69	11.00
Sri Lanka	190.43	678.44	225.45	795.99	18.39	17.33

Source: Compiled from Office of Textiles and Apparel, U.S. Department of Commerce

In Table 1.14 fiber composition of apparel import indicate highest growth of imports from China for cotton apparel during Jan-July 2005

Table 1.14: India's competitive position (fiber-wise)

(Value in US$ million)

From	Cotton apparel		MMF apparel	
	(Jan-July 04)	(Jan-July 05)	(Jan-July 04)	(Jan-July 05)
India	837.97	1231.03	213.51	229.21
China	1235.65	3668.08	1146.93	2358.36
Bangladesh	564.45	780.07	268.71	257.069
Pakistan	449.59	511.44	47.10	39.58
Sri Lanka	408.02	539.15	261.12	250.00

Source: Compiled from Office of Textiles and Apparel, U.S. Department of Commerce

followed by India. Banladesh, SriLanka followed by Pakistan in decreasing order of growth are other countries from where imports have increased in the said periods.

The imports of MMF apparel have significantly increased from China. India, Bangladesh have almost same position for MMF apparel while imports from Sri Lanka & Pakistan have decreased in corresponding period.

Trade Restrictions on China in WTO era

The high growth in imports from China to US in the period after 1st January 2005 resulted in US resorting to imposing quotas for a period of 3 years for imports from China. According to MoU (three-year quota agreement) signed by the United States and China on November, 8, US will permit the import of textile and apparel of Chinese origin falling in 34 product categories within the annual levels agreed to between them. Under the MoU, the growth that China would ac hieve for these core categories during 2006 will be no more than 10 percent, while in 2007 and 2008, growth has been fixed at modest 12.5 and 15 percent respectively. This agreement will be for the period of three years ending December 2008 and covers 34 textile and apparel products, the import of which into the US would be limited in terms of volume.

The categories are–

14 Core categories: Cotton and MMF trousers, cotton and MMF knit shirts, underwear, woven shirts, and brassieres.

20 additional categories: Socks, sweaters, swimwear, knit fabric, wool suits, wool and ramie trousers, sewing threads, combed cotton yarn, cotton towels, polyester filament, synthetic filament fiberglass and industrial fabric, textile blinds etc.

1.4.2 EU Market

The trends in EU market (Table 1.15) for Jan-June 2005 over corresponding period in 2004 indicate a minor decrease (2.9 percent) in

Table 1.15: Imports of Apparel in EU (Value in euro mn)

Apparel type	Jan-July 04	Jan-July 05	% change
Knitted	10235.02	9881.72	-3.4
Woven	13204.94	12878.82	-2.4
Total	23439.96	22760.54	-2.9

Source: Compiled from EUROSTAT, CIRFS, CITH, EURATEX

imports. The negative import is seen in knitted (-3.4 percent) as well as woven (-2.4 percent) apparel.

The EU imports has increased for W/G knitted and M/B knitted shirts, Jerseys in Jan-July 2005 over corresponding period in year 2004.The imports of categories i.e. M/B knitted suits, W/G knitted suits, T-shirts have decreased in this period.

The imports from India has increased by 14.12 percent (qty) and 14.53 (value) in Jan-July 2004-05. The growth in imports is 19.71 percent (qty) for knitted and 9.84 percent (value) while there is growth of 18.53 percent (value) for woven apparel imports in EU during this period. Import of knitted apparel has increased in quantity while woven apparel have better growth in EU imports in value.

The imports from China have increased by 52.03 percent followed by Bangladesh (26.5 percent) and India (14.12 percent). The imports from Sri Lanka have increased marginally (5 percent) while imports from Pakistan have decreased by 5.2 percent during Jan-July 2005/ 2004 in quantity (mn.kg), (Table 1.16).

Table 1.16: India's competitive position
(Qty. mn kg)

	Apparel type	Jan-July 04	Jan-July 05	% change
Bangladesh	Knitted	37.17	59.30	59.6
	Woven	70.34	76.70	9.1
		107.51	136.00	26.5
China	Knitted	212.92	312.77	46.9
	Woven	300.52	467.81	55.7
		513.44	780.58	52.03
Pakistan	Knitted	11.94	11.30	-5.3
	Woven	7.85	7.46	-5.0
		19.79	18.76	-5.2
Sri lanka	Knitted	5.83	5.59	-4.1
	Woven	12.19	13.34	9.4
	Total	18.02	18.93	5.0

Source: Compiled from EUROSTAT, CIRFS, CITH, EURATEX

Trade Restrictions on China in WTO era

The trade between China and EU has witnessed a steep increase in the period after January 2005 with imports reaching to new heights. This development made EU to look for an agreement with China. EU and China has reac hed to agreement on growth of Chinese textile imports to EU and China has reached to agreement on growth of Chine textile imports to EU till 2008. Chinese textile exports to the EU in 10 textile product categories would be limited to agreed growth levels between 8 percent and 12.5 percent till the end of 2007. The details of restricted categories along with agreed growth rates for agreement period are as follows–

Table 1.17: Restricted categories for China

Product Category	Agreed growth rate (%) for '05	Overall Quantitative limit for full yr '05	Agreed growth rate (%) for '06	Quantitative limit for full yr '06	Agreed growth rate (%) for '06	Quantitiative limit for full yr '07
Cotton fabrics (2)	12.5	55065	12.5	61948	12.5	69692
T-shirts (4)	10.0	491095	10	540024	10	594225
Pullovers (5)	8.0	181549	10	199704	10	219674
Men's trousers (6)	8.0	316430	10	348072	10	382880
Blouses (7)	8.0	73176	10	80493	10	88543
Bed linen (20)	12.5	14040	12.5	15795	12.5	17770
Dresses (26)	10.0	24547	10	27001	10	29701
Brassieres (31)	10.0	205174	10	2256921	10	248261
Table linen (39)	12.5	10977	12.5	12349	12.5	13892
Flax yarn (115)	10.0	4309	10	4740	10	5214

unit '000' pieces except for categories, 2, 20, 39 and 115 where unit is 'tons'.

The agreement thus reached between the EU and China is expected to, on the one hand, provide a three-year breathing space to the European industry, which has a huge capacity for inovation and adjustment-to get adjusted to the situation and on the other, benefit china.

SUMMARY

Indian apparel and textile sector is very critical to the Indian economy. The share of India is world apparel and textile trade is increasing. World Trade in apparel and textile has been regulated under various trade arrangements. The 'Uruguay round negotiations' has resulted in ten year phase out of quotas under ATC posing a great opportunity and challenge to Indian Industry. WTO era may see rapid increase in demand from low cost supplying countries for basic items and niche and high quality apparel producing countries for fashion or high end apparel and textiles. Trends in WTO era (post 1st Jan 2005) have indicated sharp gains for China forcing US, EU to take steps to restrict imports from China. Besides it, India, Bangladesh, Srilanka and Pakistan have also benefited from phasing out of quotas.

Chapter 2

Indian Apparel & Textile Industry

2.1. OVERVIEW

Textile exports have emerged as the largest net foreign exchange earner for the country, contributing around 21 percent of India's total export earnings. The textile exports in rupee terms have grown at an annualised growth rate of 8.52 percent during the last five years.

Textile and clothing consist of 16.11 percent of total exports of India. The share of textile and clothing in total export was 28.31 in year 1998-1999 and has always contributed significantly in Indian export (Table 2.1).

In 1998-99, the exports of all commodities from India was Rs. 139751.77 crore which has increased to Rs. 356068.88 crore in 2004-05. The export of textile and clothing in 1998-99 has increased from Rs. 39565.49 crore to Rs. 57377.66 crore in year 2004-05.

Table 2.2 shows the further distribution (fiber composition

wise) of textile and clothing in total export (net foreign exchange earnings in value) from India. As per the data of 2003-2004, cotton and 100 percent non cotton have around 13 percent share each and blended

Table 2.1: Share of Exports of Textiles and Clothing in Total Exports

Year	Exports of all commodities (Rs. Crore)	Exports of textiles & clothing #(Rs. Crore)	Percentage share of textiles/clothing in Total Exports
1998-99	139751.77	39565.49	28.31
1999-00	159095.20	45036.57	28.31
2000-01	202509.76	54638.68	26.98
2001-02	207745.56	50474.16	24.30
2002-03	252789.97	59430.08	23.51
2003-04*	293366.75	60436.16	20.60
2004-05	356068.88	57377.66	16.11

: Including Silk, Jute and Handicrafts
* : Excluding CARPET(EXCL. SILK) MILLMADE 2003-04 onwards
Source: Compiled from Compendium of Textile Statistics

Table 2.2: Fiberwise (percentage) Contribution

Year	Cotton	100% Non-Cotton	Blended/Mixed	Total
1991-92	13.71	6.26	2.90	22.87
1992-93	15.57	6.36	2.57	24.50
1993-94	15.92	6.72	3.58	26.22
1994-95	15.24	7.47	3.27	25.98
1995-96	16.32	8.19	3.48	27.99
1996-97	16.24	9.08	3.98	29.30
1997-98	15.94	10.4	14.57	30.92
1998-99	13.07	10.99	4.13	28.19
1999-00	14.16	11.91	4.48	30.55
2000-01	14.22	11.96	4.50	30.68
2001-02	14.82	12.4	4.69	31.97
2002-03	14.40	12.59	4.38	31.37
2003-04 (P)	13.41	13.09	4.51	31.01
2004-05 (P)	–	–	–	31.89

Note: The percentage is based upon 'Net foreign exchange earnings in total export'
Source: Compiled from Compendium of Textile Statistics

is having around 4.5 percent share in export. The share of cotton has been stagnant to this level in last one decade. While the percentage of 100 percent non-cotton has increased from 6.26 percent to 13 percent. The share of blended has increased from 2.9 percent to 4.51 percent between 1991-2004. The total share of exports of textile &clothing in total exports has increased from 22.87 to 31.89 percent in corresponding period.

2.2. COMPOSITION

2.2.1 Growth of Industry

The textile and apparel industry is classified into several segments, based on the type of establishment or by the fibre mix. These include: composite mills (also called the organized sector), powerloom, handloom, wool, silk and handicraft (khadi). The number of spinning mills has increased many folds in last three decades and present capacity is close to 34.24 million spindles (Table 2.3). The number of spinning mills has increased from 379 in year 1971 to 777 in year 1991 and to 1566 in year 2005 showing a remarkable increase in terms of number of spinning mills. The number of composite mills has reduced from 291 to 223 during 1971-2005. The number of spindles has increased from 18.11 million (year 1971) to 26.67 (year 1991) to 34.24 (year 2005), which is almost 100 percent increase in the spinning capacity. On the other side, although the number of composite mills has remained stagnant but number of looms in organized sector have reduced from 207800 to 86000 in last two decades showing a reduction in weaving capacity. It can be interpreted that more of the spinning mills have come and the textile trade has become more concentrated on spinning and has less emphasis on value addition i.e. weaving in last few years.

Table 2.3: Capacity in Indian textile industry

	1971	1981	1991	2001	2002	2003	2004	2005(p)
No. of mills	670	693	1062	1846	1860	1875	1887	1889
Spinning	379	415	777	1565	1579	1599	1564	1566
Composite	291	278	285	281	281	276	223	223
Spindles (mn)	18.11	21.23	26.67	35.53	35.75	36.10	34.02	34.24
Rotors (000)	—	5	66.92	394	409	379	383	385
Looms (000)	207.8	207.9	177.8	123	123	119	88	86

Source: Compiled from Compendium of Textile Statistics

2.2.2 Fibre Mix

The Indian textile industry is pre-dominantly cotton based, nearly 60 percent of overall consumption in textiles and more than 75 percent in spinning mills are of cotton. The production of raw cotton is much higher than production of man - made fiber (Table 2.4). The production of man-made fibre has increased in last few years i.e. from 114.95 million kg. (year 1980-81) to 953.33 millions kg (year 2003-04) and the production of raw cotton has reached from 1326 million kg. to 3009 million kg. between 1980-2004.

Table 2.4: Production of fibre in India (million Kg)

Fibres	1980-81	1990-91	2000-01	2003-2004
Raw Cotton	1326.0	1659.03	2380	3009.0(P)
Manmade Fibre	114.95	337.86	904.28	953.33
Raw Wool	–	36.41	47	50.7(A)
Raw Silk	5.04	12.56	15.86	15.74(P)

Source: Compiled from Compendium of Textile Statistics

During 2000-05, the production of man-made fibers has increased from 904.28 to 1018 mn. kg. The production of VSF, ASF, PSF & PPSF has almost been stagnant in the corresponding period which is shown in Table 2.5. The production of VSF has increased from 236.17 mn. kg to 245.86 million while the production of acrylic staple fiber has increased from 99.43 to 125.74 million kg. The production of PSF has increased frm 566.42 million to 643.53 million kg while the production of poly propylene staple fibre has increased from 2.26 to 2.28 million kg during 2000-05.

Table 2.5: Production of Man-made Fibres (mn. kg.)

Year	Viscose Staple Fibre	Acrylic Staple Fibre	Polyester Fibre	Poly Propylene Staple Fibre	Total
2000-01	236.17	99.43	566.42	2.26	904.28
2001-02	185.28	94.84	551.42	2.38	833.92
2002-03	224.61	105.275	82.13	2.46	914.47
2003-04	221.01	117.00	612.58	2.74	953.33
2004-05 (P)	245.86	125.74	643.53	2.88	1018.01

Source: Compiled from Compendium of Textile Statistics

As shown in table 2.6 below the consumption of MMF fibers has increased from 889 to 955 mn. Kg. during 2000-05. Table 2.5-2.6 indicates that the production and consumption of MMF fibers is almost same in corresponding period. The consumption of VSF has increased from 221 to 225 mn. kg. While consumption of PSF has increased from 562 to 610 mn. kg. The consumption of acrylic staple-fiber has increased from 104 to 118 million kg during 2000-05.

Table 2.6: Consumption of MMF Fibres

(mn. kg.)

Fibre	2000-01	2001-02	2002-03	2003-04	2004-05
Viscose Staple Fibre	221	191	225	226	225
Polyester Staple Fibre	562	556	572	596	610
Acrylic Staple Fibre	104	114	115	119	118
Other Miscellaneous	2	2	3	3	2
Total	889	863	915	944	955

Source: Compiled from Compendium of Textile Statistics

(i) Cotton

India ranks third in global cotton production after US and China. India accounts for approximately 21 percent of world's total cotton area and 16 percent of global cotton production. China and US followed by India, Pakistan are key producers of cotton(Table 2.7). Except India, all of them have clean and uniform variety of cotton. In terms of yield, Australia is world leader followed by China, Egypt and US. India is having rather poor yield of cotton (303 kg per hectare) alongwith the fact that it has more of unclean and variable characteristics in cotton.

Table 2.7: International cotton production – 2004

Country	Acreage (000 hectares)	Production (000 bales)	Yield (kg/hectare)	Characteristics
China	5,000	27,000	1,176	Clean, uniform
US	5,079	16,600	712	Clean, uniform
India	8,400	11,700	303	Unclean, variable
Pakistan	3,000	8500	617	Clean, uniform
Australia	220	1500	1484	Clean, uniform
Egypt	250	1050	953	Clean, uniform

Source: Compiled from Compendium of International Textile Statistics

Table 2.8: Cotton availability ("000 Bales of 170 Kg each)

Variety	Staple length	1991	1993	1994	1995	2000
Superior long staple	>27mm	2,929	4,284	4,010	4,899	5,970
Superiormedium staple	22-24mm	4,918	6,099	5,640	5,960	6,870
Short staple	<19mm	883	1,200	1,066	1,250	2,439
Total		9836	11,583	10,716	12,109	15,279

Source: Compiled from ICMF

Out of the total 152-lakh bales production of cotton in 2000-01, production of long and extra long staple varieties cotton together stood at 128.4 lakh bales. The production of superior medium staple (22-24mm) cotton is 68.70 lakh bales while superior long staple (>27mm) is only 59.7 lakh bales (Table 2.8). There is relatively high growth in production of short staple (<19 mm) cotton in last one decade. The percentage increase is of 176.22 while the production of superior medium staple (22-24 mm) has increased by 39.69 percent during the same period. It reflects more availability of shorter staple cotton and lesser availability of final staple variety. Among the cotton growing states Andhra Pradesh takes the lead followed by Gujarat, Maharashtra, Madhya Pradesh, Haryana, Rajasthan, Karnataka and Tamil Nadu.

(ii) Wool

India's wool industry is principally located in the northern states of Punjab, Haryana and Rajasthan with more than 75 percent of the total production capacity. The sector consists of Composite Mills, Combing units, Worsted and non-worsted spinning units and machine made carpets manufacturing units and the decentralized players- Hosiery and Knitting, Powerloom, Handlooms, and Hand Knotted Carpets and Independent dyeing processing houses. In all there are more than 700 registered units in the sector and more than 7000 powerlooms and other unorganized units. The large players in the sector have made significant inroads into the world market with supply tie-ups and joint ventures with important brands in EU and other developed countries. India depends upon imports of fine quality wool required by the organized mill and to a lesser extent the decentralized hosiery sector. Imports have been mainly from Australia and New Zealand, the major supplier is Australia. New Zealand wool is being imported mainly for Carpet sector for blending it with indigenous wool.

(iiii) Silk

India is the second largest producer of silk, contributing about 18 per cent to the world production. In India the industry is still a cottage industry. There are 4½ million Indians in 45,000 villages producing silk, with almost 200,000 handlooms and 30,000 powers looms producing 7,000 tonnes of raw silk from 200,000 ha of mulberry bushes. India also produces coarser silk from wild silkworms (Tusseh, Eri and Muga silks). Seventy percent of India's silk is produced in Karnataka State. India does not export raw silk but spins, weaves and dyes the silk to make saris, ties, upholstery, cushion covers and bed spreads for local and export sales. The strength of this industry lies in its wide base; the sustaining market demand pulls especially from the Indian handloom-weaving sector, the infrastructure created by the national sericulture project and the research and training capabilities.

(iv) Manmade fibre and filament yarn(s)

Manmade fibre and yarn segments have an equal share of volumes. Polyester fibre and Polyester filament yarn are the major products in the segment, accounting for more than 75 percent of the production. The manmade fibre industry consists of less than 100 players, all being medium and large units. In the past five years imports have shown a generally declining trend. India also exports man-made fibres and yarns and export volumes have been higher than imports in the last two years. Indian industry majorly manufactures and consumes Polyester, viscose and acrylic. India also has a strong production base for synthetic and regenerated fibres.

2.2.3 The Spinning Industry

The Indian spinning industry is dominated by cotton yarn, which also accounts for 80 percent of total value of yarn exports. The cotton yarn

Table 2.9: Production of Spun Yarn (mn. kg.)

Year	Cotton Yarn	Blended Yarn	100% Non-Cotton Yarn	Total Yarn
2000-01	2267	646	248	3161
2001-02	2212	609	280	3101
2002-03	2177	585	320	3082
2003-04	2121	589	341	3051
2004-05 (P)	2140	589	383	3112

Source: Compiled from Compendium of Textile Statistics

production has increased by 98.78 percent from 1067 mn. Kgs. to 2140 mn. kg in last two decades. The total yarn production has increased from 1298 mn. kg. in 1980–81 to 3112 mn. Kg. in 2004-2005. Table 2.9 also indicates increasing production of synthetic yarn in India in last decade. The production of spun yarn has decreased from 3161 to 3112 mn. Kg. During 2000-05. The production of cotton and blended yarn has decreased while that of 100 percent non-cotton has increased marginally from 248 to 383 mn.kg. The production of cotton yarn has reduced from 2267 to 2140 mn.kg. and that of blended yarn 646 to589 mn.kg in corresponding period.

Table 2.10: Production of Man-made Filament Yarn (mn. kgs.)

Year	Viscos Filament Yarn	Nylon Yarn Filament Yarn	Polyester Yarn Filament Yarn	Poly Propy-leneFilament Yarn	Total
2000-01	55.26	26.27	819.70	18.49	919.72
2001-02	48.35	27.82	866.16	19.84	962.17
2002-03	50.79	29.73	995.37	24.41	1100.30
2003-04	53.17	30.99	1013.00	20.82	1117.98
2004-05 (P)	53.10	37.72	998.87	16.30	1105.99

Source: Compiled from Compendium of Textile Statistics

The production of MMF yarn has increased from 920 to 1106 mn. kg. during 2000-05 (Table 2.10). The highest contribution is of Polyester filament yarn (999 mn. Kg.) followed by smaller quantity of Viscose, Nylon, Polypropylene filament yarn. (Table 2.11) shows an increase of 40 percent for Nylon filament yarn while in absolute value polyester filament yarn is still highest amongst all categories. The

Table 2.11: Consumption of Filament Yarn (mn. kg.)

Filaments	2002-03	2003-04	2004-05 (P)	% Change
Polyester Filament Yarn	992.83	1065.81	989.60	-7.15
Viscose Filament Yarn	44.52	50.80	55.10	8.50
Nylon Filament Yarn	28.34	25.97	36.48	40.47
Polypropylene Filament Yarn	22.82	21.30	14.20	-33.33
Total	1088.51	1163.88	1095.38	-5.89

Source: Compiled from Compendium of Textile Statistics

consumption of filament yarn has remained almost stagnant from 2002-05. Out of total consumption of filament yarn (1095.38 mn.sq.mt) PFY contribute 988.60 mn.sq.mts. followed by smaller quantities of Viscose, Nylon and Polypropylene filament yarn.

Table 2.12 shows that the production of coarser yarns i.e. below 20^s has increased significantly from 415 mn kg (1980-81) to 838 mn kg (2003-04) while the production of 21-40^s yarn has increased from 533 mn kg to 1015 mn kg during the similar period. The production of finer counts (above 60^s) is comparatively very low. It can be inferred that most of our spinning capacity produces Medium and low count yarn (below 40^s). India has average production of count 1-40^s.

Table 2.12: Count-wise production of cotton yarn (mn. kg.)

Count Group	1980-81	1985-86	1990-91	1995-96	1999-00	2001-02	2002-03	2003-04
1-20s	415	451	551	725	1013	963	904	838
21s-40s	533	595	733	881	979	1004	1009	1015
41-60s	67	121	129	153	131	147	161	161
61-80s	34	61	64	95	44	61	61	64
81s- and above	18	25	33	40	37	37	42	43
Total	1067	1253	1510	1894	2204	2212	2177	2121

Source: Compiled from Compendium of Textile Statistics

2.2.4 The Weaving & Knitting Industry

The fabric production industry can be divided into 3 sectors viz. Powerloom, handloom and mill sector. The decentralized sector accounts for around 95 percent of the total cloth production. The knitted fabric forms 17 percent of the total fabric production. Despite the largest loomage in the world, the weaving sector in India is a virtual mixed bag. On one hand, we have handlooms producing only 5 mtrs every 8 hrs, compared to the state-of-the-art looms producing upto 200-300 mtrs per shift of eight hours. The knitting industry is concentrated primarily in the unorganized sector with only a handful of large organized players. Knitting is primarily concentrated in cities of Tirupur and Ludhiana located in southern and northern India respectively. Tirupur accounts for nearly three-fourths of the exports of knits and specializes in cotton knits. The city of Ludhiana on the other hand caters to the domestic demand.

Table 2.13 indicate that the total number of looms in India (weaving capacity has increased by 46.20 percent during 1991-2004. Out of this, the number of looms in mill sector has decreased by 50.56 percent while the number of looms in unorganized sector has increased by 61.33 percent during 1991-2004. The number of looms in mill sector has reduced from 178000 to 88000 in that period while the number of looms in unorganized sector has increased from 1138000 to 1836000 during 1991-2004. The analysis reflect that share of organized sector in terms of weaving capacity has reduced from 13.52 percent to 4.57 percent of total number of looms while the share of unorganized sector in terms of weaving capacity has increased from 86.5 percent to 95.43 percent during 1991-2004.

Table 2.13: India's weaving capacity ('000)

Year	No. of Looms (Organized sector)	No. of Looms (Unorganized sector)	Total No. of Looms
1991	178 (13.52%)	1138 (86.5%)	1316
1995	139 (9%)	1415 (91%)	1554
2000	123 (7%)	1630 (93%)	1753
2001	123 (6.9%)	1656 (93.1%)	1779
2002	123 (6.7%)	1695 (93.3 %)	1818
2003	119 (6.57)	1692 (93.43%)	1811
2004	88 (4.57%)	1836 (95.43 %)	1924
% Change 1991/2004	-50.56	61.33	46.20

Source: Compiled from Compendium of Textile Statistics

The total number of installed capacity of looms(Table 2.14) in non-SSI sector is 103281.Out of which 85762 looms are installed in

Table 2.14: Installed Capacity of Looms in Non-SSI Sector

Item	Non-Auto	Auto	Shuttleless	Semi-Auto	Tape-Narrow Width	Total
Composite Mills	56114	19582	9631	435	Nil	85762
Exclusive Weaving Mills	10536	233	44422	16	213	17519
Total	66650	21916	14059	451	213	103281

Source: Compiled from Compendium of Textile Statistics

composite mills while 17519 are installed in exclusive weaving mills. Further to it, Out of the total looms non-auto looms are 66650 while 14059 are only shuttle less looms indicating lower percentage of technology looms in India.

2.2.4.1 Trend of Fabric Production

The manufacturing industries in India is divided into two sectors – organized and the unorganized, also sometimes referred to as the centralized and decentralized sectors. The distinction basically stems from the fact whether a particular unit is registered as an industrial unit under the relevant law or not. Any industrial undertaking employing 10 or more workers and using power or employing 20 or more workers and not using power is covered under the Factories Act and is required to be registered with the appropriate authority. This sector is subject to statutory regulations and control and may be termed as the organized sector. All industrial enterprises, which are not covered under the 'Factories act' constitute the unregistered or unorganized sector. Some of these units may be registered as small-scale industry (SSI) with the Directorate of Industries/District Industries Centre (DIC) but are relatively free from statutory regulations and control by Governmental authorities.

Powerloom sector contributes 59 percent of total fabric production followed by 19 percent by the handloom sector, 17 percent by the knit (hosiery) yarn sector and the remaining by the organized mill sector. The large share of power looms (an intermediate category of looms, operated by power) has resulted from a government policy that supports the unorganized sector in the form of reservation of product categories, mandatory export entitlement quotas and input pricing interventions. The woven fabric accounts for around 45 percent of the fabric requirement for the garment exports, whereas the rest 55 percent is accounted by the knitted sector. The woven fabric requirement is growing at a rate of around 8.5 percent whereas knitted fabric requirement is growing at a rate of 10 percent over the last 5 years. Table 2.15 shows that overall production has increased from 12444 mn. sq. mts. (year 1980-81) to 44687 mn. sq. mts. (year 2004-05) showing an increase of 259.10 percent in last two decades. Out of this, the major increase has come from powerloom sector (494 percent) while the production in mill sector has reduced from 4533 mn.sq.mts. (Year 1980-81) to 1487 (year 2004-05) showing a downward change of 67.19 percent. It indicates increasing reliance on powerloom sector while the mill sector of Indian textile and apparel industry has witnessed a downward shift in production.

Table 2.15: Sector-wise production of cloth (mn. sq.mtrs)

Year	Mill	Hand-loom	Power-loom	Hosiery	Khadi, Wool and Silk	Total
1980-81	4533	3109	4802	—	—	12444
1985-86	3544	4135	9534	—	—	17213
1990-91	2589	4295	13348	2696	402	23330
1995-96	2019	7202	17201	5038	498	31958
2000-01	1670	7506	23803	6696	558	40233
2001-02	1546	7585	25192	7067	644	42034
2002-03	1496	5980	26109	7881	662	42128
2003-04	1434	5494	26947	7847	662	42384
2004-05 (P)	1487	5676	28526	—	—	44687

Source: Compiled from Compendium of Textile Statistics

There is decrease in production of mill-sector fabric while in unorganized sector there is an increase in production from 1980-81 to 2004-05. The analysis indicates that mill sector accounts for 3.6 percent of total fabric production while power loom and handloom (unorganized) commands 96.4 percent share of total fabric production. The further analysis reflects heavy dependence of Indian textile industry on un-organized sector. It seems that the mill-sector has seen reduction in its production due to unfavorable policies and more emphasis on development of small scale or rather unorganized industry.

Table 2.16: Fibrewise Production of Cloth (mn.sq.mtr)

Year	Cotton	Blended	100% Non-Cotton	Khadi, Wool & Silk	Total
1980-81	8368	1270	1350	—	10988
1985-86	12467	1660	3086	—	17213
1990-91	15431	2371	5126	402	23330
1995-96	18900	4025	8535	498	31958
1999-00	18989	5913	13725	581	39208
2000-01	19718	6351	13606	558	40233
2001-02	19769	6287	15334	644	42034
2002-03	19300	5876	16135	662	41973
2003-04	18041	6068	17613	662	42383
2004-05 (P)	20488	6000	18200	—	45349

Source: Compiled from Compendium of Textile Statistics

Table 2.16 shows the details of production of cloth (fibre-wise). It indicates the production of cotton cloth has increased from 8368 mn.sq.mt. (year 1980-81) to 20488 mn.sq.mt. in (year 2004-05) while the change is very high for 100 percent synthetic and blended cloth. The production of cotton fabric has increased from 15431 mn sqmt (1990-91) to 20488 mn sqmt (2004-05) while the production of blended fabric has increased from 2371 mn sqmt (1990-91) to 6000 mn sqmt(2004-05) . The production of 100%non-cotton fabric has increased from 5126 mn.sq.mtr. (1990-91) to 18200 mn.sq. mt.(2004-2005). The overall cloth production has increased by 94.38 percent during same period.

The analysis of production of cloth in different sectors shows that for 100 percent synthetic (non-cotton) there is a remarkable increase in production in mill sector, while the production of cotton and blended fabric in mills sector has reduced in last two decades. The growth is rather negative for production of cloth in mill sector in last one decade. While the production of decentralized powerloom sector has increased by 88.73 percent in last decade. Out of which there is an increase in production for blended as well as 100 percent non-cotton fabric. The growth of production of cotton cloth is negative here also. The production of decentralized hosiery sector has increased in last one decade. The growth of blended fabric is highest here too.

The composition of India's fabric production (Table 2.17) indicates more than 80 percent woven fabric (volume wise) in total production of the fabric. The composition of trade in terms of woven and knitted is fairly constant in last few years. In year 1990-91, the woven fabric comprised of 89 percent of total fabric production while in year 2001-2002 the share of woven fabric was 83 percent of total production. During the similar period the percentage share of knitted fabric in total production has increased from 11 percent to 17 percent.

Table 2.17: India's fabric production (% of volume)

Year	Woven	Knitted
1990-91	89	11
1995-96	84	16
1996-97	83	17
1998-99	84	16
1999-2000	83	17
2000-2001	83	17
2001-2002	83	17

Source: Compiled from Compendium of Textile Statistics

2.2.4.2 Trend of Fabric Exports

Export of fabric from India (Table 2.18) has increased from Rs. 635707.5 lakhs to Rs. 1147048 lakhs from 1995-2004. The share of MMF in total fabric exports has increased from 27.16 percent to 38.57 percent (value terms) while for the cotton fabric for the similar period the cotton fabric exports has decreased from 57.75 percent to 38.27 percent showing a shift towards exports of blended or synthetic fabrics. The trends of fabric exports for a period of 1995-2004 indicates an increase (80.43 percent) in the fabric exports in value terms. The increase is more for MMF, silk and special woven categories. The growth is rather sluggish for exports of cotton fabrics.

Table 2.18: Trends of fabric exports (Value in Rs. Lakh)

FABRICS	1995-96		2000-2001		2003-04	
	% Value	Value	% Value	Value	% Value	Value
Silk	10.27	65285.8	13.06	123008.8	13.21	151574
Woollen	2.39	15221	1.44	13551.55		7877
Cotton	57.75	367111	47.33	445696	38.27	439019
MMF	27.16	172654	24.03	226285.9	38.57	442493
Flax&Jute	1.00	6374.5	2.96	27897.23	2.72	31279
Pile&knitted	1.43	9061.2	1.62	15289.18	2.06	23730
Spl.woven fabrics	10.27	76889.36	9.55	89909.13	4.45	51077
Total	100	635707.5	100	941637.8	100	1147048

Source: Compiled from Compendium of textile statistics

Table 2.19 of composition of fabric (i.e. Woven vs. Knitted) exported indicates that woven consists of 96.2 percent of total export of fabric. It is also evident that export pattern is almost stagnant in last few years. In year 1995-96 the woven consisted 99.35 percent of total export, while in year 2000-01 the woven fabric consists 96.2 percent of total fabric exports. The export of knitted fabric has increased from 0.65 percent to 3.8 percent of total fabric export.

Table 2.19: India's fabric export (mn.sq.mts.)

Year	Woven	Knitted
1995-96	2315 (99.35%)	15 (0.65%)
1996-97	2407 (94.8%)	131(5.2%)
1998-99	2700 (96.1%)	107 (3.9 %)
1999-2000	2708 (96.6%)	95 (3.4%)
2000-2001	2750 (96.5%)	98 (3.5%)
2001-02	2834 (96.2%)	112(3.8%)
% Change 2001-02/1995-96	22.41 %	646.67%

Source: Compiled from Compendium of Textile Statistics

The analysis of process wise exports (Table 2.20) of cotton fabric indicates that Grey (33.67 percent), Yarn dyed (26.96 percent), Piece dyed (10.13 percent)and bleached (9.54 percent) are major categories of total exports of cotton fabric. The similar ranking was prevalent during last year.Grey, Yarn dyed followed by Piece dyed, bleached and printed cotton fabrics are leading in quantity terms too in exports.

Table 2.20: Processwise Exports of Cotton Fabrics

Process	Apr 2004/ Mar 2005				
	Qty		Value		
	Mn.Kgs	Mn.Rs	Mn.US$	Rs/Kg	%age
Grey	242.59	14020.27	312.33	57.79	33.67
Yarn Dyed	147.83	11228.74	250.14	75.96	26.96
Piece Dyed	69.66	4220.08	94.01	60.58	10.13
Bleached	71.13	3972.85	88.50	55.85	9.54
Printed	55.69	2880.28	64.16	51.72	6.92
Embd. Cloth	17.14	1309.83	29.18	76.42	3.15
Zari Work	11.78	950.25	21.17	80.67	2.28
Hand Ptd	4.39	417.49	9.30	95.10	1.00
Hand Dyed	0.29	115.13	2.56	397.00	0.28
N E S	23.60	2530.76	56.38	107.24	6.08
TOTAL	644.10	41645.68	927.73	64.66	100.00

Source: Compiled from TEXPROCIL/DGCIS, Kolkata

2.2.4.3 Trend of Fabric Import

The import of synthetic fabric is dominating the import basket for fabric and is having a share of 27.22 percent of total fabric imports in 2003-2004 (Table 2.21) while share of import of cotton fabric is 25.45 percent of total fabric imports in 2003-2004. There is a remarkable increase (103 percent) in fabric imports from 1995-2004, while the shift is marginal for MMF. The import of MMF fabric contributes 27.22 percent of total fabric import in 2003-2004 while in year 1995-96 the percentage of import of MMF fabric was 7.19 percent. There is an increase in import of pile and knitted fabric in value terms during the same period and the share of pile and knitted fabric in total fabric import stands 8.48 percent in 2003-2004 while it was 9.46 percent in 1995-96. The increase in import is evident in silk, MMF and knitted and special woven fabric category.

Table 2.21: Trends of fabric imports

(Value in Rs. Lakh)

Fabrics	1995-96		2000-2001		2003-04	
	% value	Value	% value	Value	% value	Value
Silk	3.86	463.86	2.70	2054.1	16.51	40223
Woollen	1.78	214.95	0.99	751.36	6.28	15312
Cotton	31.71	3839.04	14.17	10788.46	25.45	62001
MMF	7.19	851.8	5.25	4001.73	27.22	66329
Flax Jute	34.64	4198.5	34.15	26004.2	5.80	14150
Pile knitted	9.46	1139.71	29.96	22813.48	8.48	20667
Spl.woven fabrics	11.36	1319.67	12.79	9743.62	10.23	24929
Total	100	11997.0	100	76156.95	100	243612

Source: Compiled from Compendium of Textile Statistics

The analysis of major countries for imports of MMF fabric for India (Table 2.22) indicates that China followed by Korea & Taipaei are leading source of fabricator Indian industry. The other countries include Nepal, Italy, Indonesia etc. The value of imports from China has increased from Rs. 2945.78 to 20616 Lakhs during 2001-04. While imports from Korea has increased from Rs 8450 to 15263 Lakhs. The total import of fabric has increased from Rs. 26004 to 66328 Lakhs in corresponding period.

Table 2.22: Leading exporters of man made fabrics to India

(qty in tones) (value in Rs. lakh)

Country	2000-01		2003-04	
	Qty	*Value*	*Qty*	*Value*
China	1630.8	2945.78	35877.8	20616.9
Korea	1442.2	8450.5	23160.4	15263.6
Chinese Taipei	2435.5	4426.8	17186.3	13156.5
Nepal	1871.3	1097.5	3552.5	2663.03
Italy	186.3	524.9	1507.8	2337.02
Other countries	10546.36	8558.72	17227.99	12291.82
Total	18112.46	26004.2	98512.79	66328.87

Note: qty for the year 2003-04 is in thousand sq mtr.

Source: Compiled from Compendium of Textile Statistics, DGCI&S Kolkata

The import of cotton fabric (Table 2.23) has increased from Rs. 10788 to Rs. 62001 Lakhs during 2000-04. The import from China has increased from Rs. 2172 to Rs. 22062 Lakhs and from Hongkong (Rs.985 to 10203 Lakh), Japan (Rs. 597 to 5227 Lakhs). The other countries Taipei, Italy, Korea are also source of importing cotton fabric.

Table 2.23: Leading exporters of cotton fabrics to India

(qty in tones)(value in Rs. lakh)

Sl no	Country	2000-01		2003-04	
		Qty	*Value*	*Qty*	*Value*
1.	China	1097.84	2172.44	21568.72	22062.45
2.	Hong Kong	456.92	985.35	8125.39	10202.49
3.	Japan	289.92	597.36	2868.42	5227.80
4.	Taipei	580.25	1382.78	3800.85	5181.38
5.	Italy	355.98	769.95	1996.85	3920.79
6.	Korea	520.06	1287.35	2648.07	3635.22
7.	Other countries	2177.73	3593.23	10369.76	11770.92
	Total	5478.70	10788.46	51378.06	62001.05

Note: qty for the year 2003-04 is in thousand sq mtr.

Source: Compiled from Compendium of Textile Statistics, DGCI&S Kolkata

Summary of the balance of India's textile trade during 1995-2004 shows that the fibre production has increased marginally from 3446.72 million (Table 2.24) kg. to 4028.44 million kg. while imports of fibre has increased from 293.45 million kg. to 488261.57 million kg. during 1995 – 2004. The exports have increased from 80.97 million kg. to 274508.73 million kg. It also indicates that the production of yarn has increased from 2485 million kg. to 4170 million kg. while imports have increased from 56.28 million kg. to 165256.44 million kg. The exports have also increased from 589.36 million kg. to 842411.66 million kg. during 1995 – 2004.

The production of fabric has increased from 31958 million sq. mt. to 42383 million sq. mt. and exports have increased from 2340 million sq. mt. to 546524.71 million sq. mt. during 1995-2003. During 1995-2003, the imports have reduced from 36.110 million sq. mt. to 20.129 million sq. mt. and again increased to 84215.63 million sq. mt. in 2002-03.

Table 2.24: Summary of India's Textile Trade

Year	Production	Imports	Exports
Fibre (Million Kg)			
1995-96	3446.72	293.45	80.97
2000-01	3449.06	383557.36	74568.40
2001-02	3588.05	591041.78	51029.95
2002-03	3293.02	518870.85	87257.75
2003-04	4028.44	488261.57	274508.73
Change2001-04	85.61	27.29	368.13
Yarn (Million Kg)			
1995-96	2485	56.28	589.36
2000-01	4080	103720.25	886699.46
2001-02	4053	138779.94	723460.20
2002-03	4181	177428.07	871168.91
2003-04	4170	165256.44	842411.66
Change2001-04	2.20	59.32	18.40
Fabric (Million sq. mtrs)			
1995-96	31958	36.110	2340
2000-01	40233	47436.57	459041.73
2001-02	42034	68395.84	414820.37
2002-03	41973	84215.63	546524.71
2003-04	42383	–	–
Change2001-04	5.34	77.53	19.05

Source: Compiled from Compendium of Textile Statistics

2.2.5 The Indian Fabric Processing Industry

Processing is the weakest link in India's entire textile chain. The processing industry is decentralized and is marked by hand processing units, independent units and the composite mill sector. The processing sector is one of the weak links in the textile supply chain. The processing industry is dominated by Hand Processing which constitutes 82.5 per cent of the total number of processing units. Power processing units can be divided into Independent Process Houses that do job work and those with composite mills that process their own fabric. Around 89 per cent of power processing units are Independent.

New units with modern processing technologies that can add value to garments such as anti-microbial and wrinkle-free finishes are being planned to face the competitive environment. In the backdrop of expected flood of imports, value-added processes are focused to offset the effect. If processing is not given adequate attention, there are opinions that exports will continue to be dominated by grey fabrics which are then processed abroad. With this logic, the Ministry of Textiles is proposing huge modernization and setting up of large modern processing houses to increase the export of processed fabrics. Upgradation in this segment is also intended to promote integrated large units with an improved quality and lower cost structure. Hand Processors will be phased out in a process. The earlier policy objective of employment generation through discrimination in favour of hand processors is replaced by modernization and skill-intensive employment. Indian processing industry has deployed low-end technology with little investment initiative in technology upgradation. The Indian processing industry lacks R&D and innovation. At present nearly 4400 processing units are existing both in independent and power processing sector. Most of the processing in India is happening in the decentralized sector. Here again, quite like in weaving, there exists the hand-processing sector. Exhibit 2.1 shows diagrammatic description of Indian processing industry.

Over the years of modernization, the processing sector has remained weak and thus the international buyers as well as the garment manufacturers, do not find the well processed, high quality, long length defect free, larger width textile material as per their needs. The Indian textile industry needs to move up in the value chain and instead of exporting the goods in grey forms. The supply in finished form or in the garment form, will surely ensure the profit margins increase in manifolds. The present scenario of Indian processing sector indicates that out of around 2500 process houses, the 4 percent fall under composite sector, 7 percent fall under semi-composite sector and 89 percent are the independent process houses, majority of which are carrying out job-work. In other words, the processing sector is highly fragmented and decentralized; it is very old and lacks modernization. It can process the goods in small quantity and thus can not meet the needs of mass scale production.

Exhibit 2.1. Diagrammatic description of the Indian processing industry

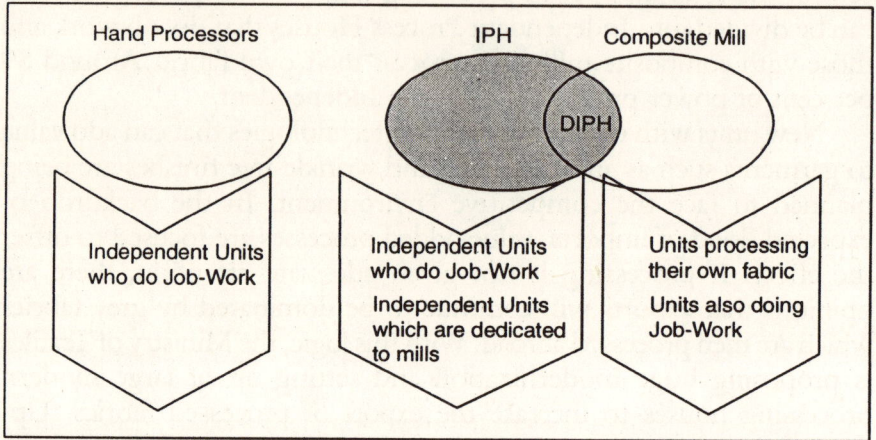

Source: www.ciionline.org

Note: The size of the figures is not representative of the size of the sector

*IPH: Independent Process House

DIPH: Dedicated Process House

2.2.6 The Indian Apparel Industry

The manufacturing of apparel has remained a labour intensive industry, offering a great advantage to low wage countries like India. The study by Kathuria and Anjali Bhardwaj (1998) classified garment manufacturing in three stages:-

1. Cutting the fabric to patterns, usually done by power-operated cutting machines; making or sewing the garment on sewing machines, either foot-operated or power-operated.
2. Sewing: The most labour-intensive part of the process is the sewing operation.
3. Finishing the garment by trimming, checking for dimensions washing, ironing and packing.

The apparel sector is structurally a labour intensive, low wage industry with some differences across its market segments. The total apparel market in India including tailored and ready-made goods is estimated to be US$ 20 billion. More than 50 percent of the Indian market is for traditional wear (sari, dhoti, Salwar, etc), which does not go into fabrication or is tailored at home. The segment is extremely fragmented with an estimated 40000 domestic manufacturers, 50,000 fabricators (job contractors) and around 1000 manufacturer-exporters. At present in India nearly 90,000 small and big organized manufacturers exists. Fabricators dominate the scene with a share of 70 percent of the estimated manufacturing capacity of 2 million machines. The apparel sector market is around US$ 9 billion of which exports account for more than US$ 5.2 billion (2003-04). Table 2.25 shows that 4.3 billion domestic market is essentially in urban areas where the consumption of ready-made apparel has risen significantly in recent years. Ready-made apparel accounts for only 20 percent of the domestic market, given the low penetration of ready-mades; most of non-urban India still depends on custom tailoring as the major source of apparel. However, brands account for nearly two-thirds of the ready-made apparel segment. Overall apparel consumption has grown at a pace of 5-6 percent in quantity terms. Men's wear is having largest share followed by women's wear and kid's wear in Indian domestic market.

Table 2.25: India's domestic Apparel market

Category	Market size $bn	Volume share%
Men's wear	1.98	45.93
Women's wear	1.59	36.89
Children's wear	0.73	16.93
Infant wear	0.23	5.33
Total	4.31	100

Source: Compiled from Images/KSA Technopak 2004

With the growth of economy in India and increase in purchasing power the market of branded apparel is increasing in past few years. The growth of branded segment is comparatively high for men's wear. The branded apparel including men's wear as well as women's wear has registered a good growth in last few years. The branded Apparel market has increased from Rs. 51.57 bn. in year 1998-1999 to Rs. 109.93 bn. in year 2003-2004 (Table 2.26).

Table 2.26: Branded Apparel market (Rs. bn)

Category	1998-99	1999-00	2000-01	2001-02	2002-03	2003-04
Men	30.67	36.50	43.80	53.45	58.34	60.45
Women	17.00	20.75	25.32	31.14	35.23	41.80
Children	3.90	4.36	4.80	5.45	6.30	7.68
Total	51.57	61.61	71.52	90.04	99.87	109.93

Source: *Compiled from Images/KSA Technopak 2004*

The apparel exports from India has increased from US$ bn 2.53 to US$ bn 5.2 during 1990-2004. During 2003-04, the exports of apparel have reached to US$ 5.2 bn as compared to US$ 5 bn in the previous year, registering a growth of 4 percent. The world apparel exports is around US$ 226 billion and the share of India is 2.58 percent. Indian ranks at position 6 after China (20.54 percent share), Hong Kong China (11.09 percent share), Turkey (4 percent share), Mexico (3.86 percent share) and US (3 percent share). The share of Bangladesh is 2.06 percent, Indonesia 1.9 percent, Korea 1.84 percent, Thailand 1.68 percent, Philippines 1.30 percent, Sri Lanka 1.16 percent and Pakistan 1.11 percent.

The average UVR of exports of apparel from India is US$ 4.01 in year 2004. It remained almost stagnant in last one decade as the UVR in 1993 was US$ 3.82, which has witnessed a marginal change but remained below US$ 4 from 1993-2004 (Table 2.27). European Union is largest destination market for export from India i.e. it constitutes 52.04 percent share of our total exports of apparel while US is single largest market with share of 40.16 percent in total exports from India. Besides it non-quota countries (UAE, Japan, Australia and Switzerland etc.) constitutes 4.16 percent of India's total exports. The average UVR from exports to non-quota countries is US$ 3.75 (2004), which is lesser than US$ 5.06 (1994). US is single largest market with highest average UVR in last one decade followed by European Union (in terms of volume) and non-quota countries (in terms of value). 52.04 percent of India's export are targeted to EU, UK (588 US$ mn), France (531.9 US$ mn), Germany (507.9 US$ mn) and Italy (214.6 US$ mn) are key markets in EU for Indian apparel exports in 2003-04. This also indicates India's target of exports of apparel is primarily low-end customer and share of value added items contributing more in average UVR is rather negligible.

Table 2.27: India's Apparel Export

Country	Year	Qnty (Mn Pcs.)	Value (Mn Us$)	Avg. UVR (Us$)	% Share (2004)
US	1993	151.6	891.7	5.9	40.16
	2003	393.7	2031.1	5.16	
	2004	392.2	1988.10	5.06	
European Union	1993	402.6	1645.2	4.09	52.04
	2003	726.5	2421.6	3.33	
	2004	732.6	2576.90	3.51	
Canada	1993	58	198	3.41	3.61
	2003	59.7	189.8	3.18	
	2004	55.7	179.20	3.21	
Non-quota Countries	1993	61.1	255.25	4.17	4.16
(UAE, Japan, Australia	2003	62.5	231	3.7	
& Switzerland)	2004	55.1	206.70	3.75	
Total	1993	905.2	3466.6	3.82	100
	2003	1242.4	4873.9	3.92	
	2004	1235.7	4950.90	4.01	

Source: Compiled from Handbook of Export Statistics, AEPC

The export of apparel from India has increased by 42.8 percent in value terms and 36.51 percent in volume terms. The average UVR has increased by 4.97 percent during 1993 to 2004 period. The percentage increase is most prominent in exports to US where, the exports have increased by 158.70 percent (volume) and 122.95 (value) in last decade. The increase in exports to EU is 81.96 percent (volume) and 56.63(value) in corresponding period. The exports to non-quota countries have decreased by 9.81 percent (volume) and 19.02 percent (value).

The apparel exports from India primarily consist of Ladies blouses, T-shirts, Gents shirts followed by Trousers/Shorts, Ladies Skirts & Jackets/coats, Ladies dresses. These categories account for around 58.4 percent of total export. The exports in these categories have increased by around 10 percent during 1998-2004 (Table 2.28). The positive percentage change is notable in Trousers, T-shirts & Jackets while in all other categories the growth is rather slow.

Table 2.28: India's Apparel export in leading quota categories

(US$ mn)

Category	1998-99	1999-2000	2000-01	2001-02	2002-03	2003-04
Gents Shirts	640.72	652.31	693.95	648.41	465.44	540.81
T-Shirts	549.08	625.85	756.82	700.12	740.01	705.64
Ladies Blouses/ Shirts	766.83	780.93	793.77	709.25	802.9	672.22
Trouser/Shorts	259.64	328.3	386.69	421.6	265.88	344.20
Ladies Dresses	288.52	254.84	286.14	198.79	229.59	241.1
Ladies Skirts	225.88	253.16	264.65	200.54	220.1	229.04
Jackets/Coats	128.91	170.66	192.84	199.72	105.29	128.68
Total	2859.6	3066.1	3374.9	3078.4	2829.21	2896.7
% of Total Export	54.35	58.28	63.42	69.79	59.31	58.40
Total Garment Export (USD bn)	5.26	5.25	5.3	4.41	4.77	4.95

Source: Compiled from Handbook of Export Statistics, AEPC

Table 2.29 indicates that average UVR of exports from India in Year 2003-2004 is US$ 4. The average UVR is maximum for Jackets/coats i.e. US$ 10.5 followed by Trousers/Shorts (US$ 8.51), Ladies dresses (US$ 5.65). The average UVR is minimum for T-shirts (i.e. US$ 3.79) & Ladies Blouses (i.e. US$ bn 4.55). The average price realization form exports are decreasing in last few years except for Trousers/shorts where there is increase in average UVR.

Table 2.29: India's UVR of Apparel Exports

(US$ per piece)

Category	1998-99	1999-2000	2000-01	2001-02	2002-03	2003-04
Gents Shirts	6.1	5.75	5.25	4.95	4.05	4.56
T- Shirts	4.3	3.76	3.5	3.41	3.68	3.79
Ladies Blouses/Shirts	4.71	3.82	3.53	3.57	4.54	4.55
Trouser/Shorts	5.1	6.45	6.8	6.95	7.84	8.51
Ladies Dresses	9.5	8.3	7.6	6.85	5.52	5.65
Ladies Skirts	5.29	5.26	4.85	4.36	5.14	5.25
Jackets/Coats	11.2	10.38	10.2	10.5	9.26	10.5
Avg. UVR	3.78	3.65	3.81	3.50	3.71	4.00

Source: Compiled from Handbook of Export Statistics, AEPC

The analysis of regionwise shipment (value) From Indian (Table 2.30) indicate that Delhi region has largest export base will 37.47 percent of export (value) followed by Mumbai (18.32 percent) & Tirupur (16.88 per cent). The quantity wise export indicate 30.62 percent shipment for Tirupur followed by Delhi (28.30 per cent) & Mumbai (10.60 percent). Tirupur ships more of average low value (per pc) export being highest in quantity. There are around 14531 registered apparel exporters with Apparel Export Promotion council (AEPC). Out of which 10508 are merchant exporters while 4023 are manufacturer exporters. The largest number of registered apparel exporters are from northern region (6531) followed by southern region (4857), western region (2766) and eastern region (377).

Table 2.30: Regionwise Exports from India

Region	Quantity	Qty %	Value in US$	Value %
Delhi	3516	28.30	18263	37.47
Mumbai	2311	18.60	8929	18.32
Kolkata	267	2.15	399	0.82
Chennai	991	7.98	4686	9.61
Bangalore	1108	8.92	6399	13.13
Jaipur	169	1.36	615	1.26
Tirupur	3804	30.62	8229	16.88
Ludhiana	259	2.08	1217	2.50
Cochin	0	0	0	0
Hyderabad	0	0	0	0
Total	12425	100	48737	100

Source: Compiled from Handbook of Export Statistics, AEPC

2.2.7 Impact of Government Policy Intervention

For the textile industry, the policy framework in India has, till recently, been very restrictive. Under the highly interventionist industrial policy regime, originating under the Industrial Policy Resolution, 1948, the textile industry was one of the eighteen 'basic' industries, whose regulation and control was considered necessary in the national interest. It was accordingly, brought within the purview of the very strict regulatory regime of the Industries (Development and Regulation) Act, 1951, which prohibited setting up of new capacities or expansion without a licence. With the nation's obsession with the Gandhian ideology, khadi and the handloom were perceived as fulfilling the twin objectives of providing employment to the rural masses and producing cheap cloth for human consumption. The government policy had a strong' anti-mill' bias to protect the handloom sector. This resulted in the government reserving in 1950 certain areas of production for exclusive manufacture by handlooms. This was followed by fiscal levies on mill made cloth in 1952, which increased in severity and scope over the years. In furtherance of this policy, weaving capacity in the mill sector was frozen at the existing level in 1956, with expansion permitted only for exports. These restrictions on textile mills also seem to have been in conformity with the development paradigm of the time provided by the Mahalanobis model, which emphasized the development of heavy industry for the achievement of the growth objective and small scale labour intensive production of essential consumer goods for providing employment.

The restraint on mill capacity had an unintended effect. It resulted in the mushroom growth of powerlooms in the unorganized sector. Even official estimates put the growth to a six-fold increase between 1951 and 1964 (from 23800 estimated by the Kanungo Committee to about 145,800 estimated by the Asoka Mehta Committee). The reservation meant for the handloom sector had actually benefited the powerloom sector. Concern for the handlooms resulted in the government reducing the disparity in incidence of excise duty between the powerloom and the mill sectors. The excise duty on hank yarn consumed by the handloom sector was completely abolished. Alongside the mill, powerlooms also began to the perceived as a threat to the handloom sector. This concern found reflection in the Government of India's textile policy of 1978, which sought to freeze powerloom capacity at existing levels. However, the proposed legislation to give effect to this policy never materialized. In 1981, the government announced a fresh textile policy, which permitted only a marginal expansion of the powerloom sector, for Handloom Cooperative Societies wishing to install powerlooms. As in the previous policy, the existing unauthorized powerlooms were to be regularized on payment of a fee. Notwithstanding the restrictive policies, the powerlooms continued to multiply. From an estimated figure of around six lakh in 1981-82, their number grew to an estimated 8.36 lakh as on 1st January 1985 taking a pragmatic view of the ground realities, the textile policy of June 1985 largely did away with the physical curbs on the growth of powerlooms. The policy provided for compulsory registration of powerlooms and asserted that mills and powerlooms would be treated on par and the cost disadvantage for the handloom vis-à-vis the powerlooms would be eliminated by appropriate adjustment in fiscal policies. The same policy also did away with the virtual freeze on weaving capacity of the mills, which had existed since 1956.

Another important concern of the textile policy has been with the provision of cheap cloth for the weaker section of society. All composite mills were required in 1964 to produce 'controlled cloth' to be sold at Government fixed prices which sometimes did not even cover the cost of production. The stipulated quantity was 45 per cent of the total mill production, later raised to 50 per cent in January 1965. The percentage was reduced to 25 per cent in 1968 and by 1971 the obligation on individual mills was replaced by an industry wide obligation. The controlled cloth scheme forced the mills to raise the prices of the non-controlled varieties; thereby adding to the poor competitiveness of the mill made cloth vis-à-vis the output of the decentralized powerloom sector·

The Textile policy of 1978 in fact recognized that the controlled cloth scheme was a major contributory factor in the spread of sickness in the organized mill sector. The policy proposed a transfer of such cloth to the handloom sector so that the employment objective could also be served while avoiding the burden on the mills. Till such time, as this could be done, the responsibility would be shouldered by the public sector National Textiles Corporation (NTC). The 1981 textile policy completely exempted the private sector from the scheme and envisaged a complete transfer of controlled cloth production from the NTC to the handloom sector by the end of the Seventh Five Year Plan (1985-90).

The Textile policy pursued contradictory objectives, again partly dictated by the Mahalanobis model, which attempted to combine essentially antithetical elements of Soviet model of development with its emphasis on comprehensive state planning and rapid industrialization with Gandhian economic belief in the efficacy of pre-industrial non mechanized techniques for the manufacture of yarn and cloth. This resulted in the retarded growth of the modern textile industry. The Mahalanobis strategy constituted a radical departure from the principle of comparative advantage, and deviated sharply from the textile led pattern of growth followed successfully by countries such as Japan. The hold of controls has also made the Indian textile industry, especially its organized sector, less competitive.

There were strict government controls on organized mills-modernization of plants, restrictions on imported machines, labour force restructuring, etc. which resulted in poor productivity, loss of competitiveness, rising operational costs, declining share of output and sickness in organized mills. In addition to it, government policies have consistently shown a bias in favor of cotton, and prevented the emergence of a world-competitive synthetic -fibre segment in India, which denies India opportunities in the world market 60 percent of which is in synthetics.

In recent years realizing the need to de-bottleneck various segment to compete in the world market the government has attempted to revitalize the textile industry through measures like delicensing, incentives for modernization, reduced import duties on textile machinery and other import-export measures, a progressive equalization of tariffs between cotton and non cotton products, and the entry of foreign direct investment in the sector. Various initiatives of government are discussed herewith–

2.2.7.1 *National Textile Policy (2000)*

The National Textile Policy-2000 was announced by Hon'ble Union Minister for Textiles on 2nd November, 2000. The highlights of the new textile policy are enumerated below:

Vision

Endowed as the Indian Textile Industry is with multifaceted advantages, it shall be the policy of the Government to develop a strong and vibrant industry that can

- Produce cloth of good quality at acceptable prices to meet the growing needs of the people;
- Increasingly contribute to the provision of sustainable employment and the economic growth of the nation; and
- Compete with confidence for an increasing share of the global market.

Objectives

The objectives of the policy are to –

- Facifitate the Textile Industry to attain and sustain a pre-eminent global standing in the manufacture and export of clothing;
- Equip the Industry to withstand pressures of import penetration and maintain a dominant presence in the domestic market;
- Liberalise controls and regulations so that the different segments of the textile industry are enabled to perform in a greater competitive environment;
- Enable the industry to build world class state-of-the-art manufacturing capabilities in conformity with environmental

standards, and for this purpose to encourage both Foreign Direct Investment as well as research and development in the sector;

- Develop a strong multi-fibre base with thrust of product upgradation and diversification;
- Sustain and strengthen the traditional knowledge, skills and capabilities of our weavers and craftspeople;
- Enrich human resource skills and capabilities, with special emphasis on those working in the decentralised sectors of the Industry; and for this purpose to revitalise the Institutional structure;
- Expand productive employment by enabling the growth of the Industry, with particular effort directed to enhancing the benefits to the north east region;

Thrust Areas

In furtherance of the objectives, the strategic thrust will be on:

- Technological upgradation
- Enhancement of Productivity
- Quality Consciousness
- Strengthening of the raw material base
- Product Diversification
- Increase in exports and innovative marketing strategies
- Financing arrangements
- Maximising employment opportunities
- Integrated Human Resource Development

Important Targets and Outputs

The endeavour will be to –

- Achieve the target of textile and apparel exports from the present level of US$ 11 billion to US$ 50 billion by 2010 of which the share of garments will be US$ 25 billion.
- Implement vigorously, in a time bound manner, the Technology Upgradation Fund Scheme (TUFS) covering all manufacturing segments of the industry;
- Assist the private sector to set up specialised financial arrangements to fund the diverse needs of the textile industry;

Powerloom Industry

The powerloom sector occupies a pivotal position in the Indian textile industry. However, its growth has been stunted by technological obsolescence, fragmented structure, low productivity and low-end quality products. The focus will therefore be on:

- Technology upgradation;
- Clustering of facilities to achieve optimum levels of production;

Knitting

Hosiery knitting, growth of which accelerated during the last decade, primarily because of expansion of hosiery into global fashion knitwear is expected to expand into the apparel and home furnishing sectors. In this segment, the following measures will be taken:

- Review of the Policy of SSI Reservation for this sector;
- Encouragement to Technology Upgradation and expansion of capacity; and
- Introduction of support systems for commercial intelligence, design and fashion inputs.

Processing and Finishing

Processing is the weakest link in the textile production chain, and results in loss of potential value. To bring about the necessary improvement.

- Government will encourage setting up of modern processing units, meeting international quality and environmental norms;

Clothing

The role of this sector is poised for radical changes in view of the changes in the international trading environment brought about by the rules and regulations of the WTO. The industry will be restructured as follows:

- The office of the Textile Commissioner will focus attention on the development of the garment industry;
- Garment industry will be taken out of the SSI reservation list;
- Joint ventures and strategic alliances with leading world manufacturers will be promoted;
- Set up a Venture Capital Fund for tapping knowledge based entrepreneurs of the industry;

- Encourage the private sector to set up world class, environment-friendly, integrated textile complexes and textile processing units in different parts of the country;
- De-reserve the Garment industry from the Small Scale Industry sector;
- Facilitate the growth and strengthen HRD Institutions
- Transform, right size and professionalise all field organisations under the Ministry of Textiles to enable them to play the role of facilitators of change and growth.

Sectoral Initiatives

Within the framework of the Policy, the following sector—specific initiatives will be taken:

Man-Made Fibre

Full fibre flexibility between cotton and man-made fibres and consumption of specialised man-made fibres/yarns will be encouraged.

Weaving sector

Despite a 58 percent global share of looms, consisting of 3.5 million handlooms and 1.8 million powerlooms, technology still remains backward. This sector, critical to the survival of the Indian textile industry and its export thrust, will be rapidly modernised. Clustering of production facilities in the decentralised sector will be encouraged to achieve optimum size and adopt appropriate technology.

Organised Mill Industry

Efforts will be made to restore the organised mill industry to its position of preeminence to meet international demand for high value, large volume products. For this purpose, the following measures will be initiated:

- Integration of production efforts on technology driven lines;
- Encouragement to setting up of large integrated textile complexes;
- Strategic alliances with international textile majors, with focus on new products and retailing strategies;
- Schemes with necessary infrastructural facilities for the establishment of textile/apparel parks will be designed with the

active involvement of State Governments, Financial Institutions and the private sector; and

- Setting up of strong domestic retail chains to ensure easy availability of branded Indian products will be encouraged.

Exports

Textile exports play a crucial role in the overall exports from India. With the objective of increasing exports to US$ 50 billion by 2010 from the present level of US$ 11 billion, the thrust will be on:

- establishing a multi-disciplinary institutional mechanism to formulate policy measures and specific action plans, including those relating to the WTO; and closely monitoring financing proposals;
- forging of strategic alliances for gaining access to technology;
- operating a brand equity fund exclusively for textile and apparel products, consistent with WTO norms.
- restructuring AEPC and other Export Promotion Councils play the role of facilitators and professional consultants;
- developing infrastructural facilities in the predominantly textile and apparel export oriented areas in close co-operation with State Governments and Financial Institutions and the private sector; and
- evolving a suitable mechanism to facilitate industry associations to deal with disputes under the various agreements of the WTO.

Other Thrust Areas

Human Resource Development

HRD assumes new significance with inescapable competition facing Indian textile products both in the international and domestic markets. Government will support programmes of organisations and institutions engaged in HRD that address the professional manpower needs of the industry, as well as at the cutting edge level of workers and shop-floor supervisors. Institutions will be encouraged to network and synergistically co-operate amongst themselves. IT will become an integral part of HRD effort.

Fiscal and Financing Arrangements

A growth-oriented fiscal road map will be drawn up, which has the advantage of predictability. The parameters within which the multi-level duty structure and rates of levies will be reviewed and rationalised will include the thrust on exports, the fiscal regime of major competing countries, WTO consistency, and the need to keep prices at levels affordable to the largely poor consumers, who will continue to form the bulk of the market.

Funding requirements of different segments of the textile industry will be periodically reviewed and short-term and long-term requirements spelled out. Innovative measures for tapping public and private sector funding will be worked out. The endeavor will be to:

- Encourage the private sector to take the initiative in participating in financing of specific needs of the textile industry;
- Set up a Venture Capital Fund in consultation with and involvement of financial institutions for the promotion of talented Indian Designers, Technologists, innovative market leaders and e-commerce ventures;

Delivery Mechanisms for Implementation of the Policy

Organisations working under the Ministry of Textiles will be re-oriented, rightsized and restructured to act as facilitators instead of regulatory bodies, with the mandate and role of each being reviewed and redefined over the next two years. Simultaneously, regulations and controls will be reviewed and progressively reduced.

Some of the specific changes will be:

- Export Promotion Councils will be restructured so as to become capable of devising dynamic export strategies; promoting financing; disseminating information on various aspects of the WTO agreements; extending legal advice to trade and industry in dispute settlements, etc.

2.2.7.2 *Technology Upgradation Fund Scheme-1999 (TUFS)*

The highlights of "Technology Upgradation Fund Scheme-1999 (TUFS)" which was launched on 01.04.1999 and extended upto 31.03.2007 are as follows –

- The scheme aims to provide funds for technology upgradation of existing units and setting up new units with state-of the-art technology.
- Interest incentive (5 perc ent) on the term loans, re-imbursement of exchange rate fluctuation (up to 5 percent) on foreign currency loans and capital subsidy (20 percent) on the small-scale power loom units up to a cost of Rs. 60 lacs are provided.

- Original outlay of Rs. 25000 crores; additional funds of Rs. 435 crores are allocated.
- Till 30.09.2004 loans aggregating to Rs. 9000 crores have been sanctioned, Rs. 7500 crores has been disbursed (project cost-Rs. 15570 crores).

2.2.7.3 *Apparel Park for Export Scheme*

The main objective is to impart focused thrust to setting up apparel manufacturing units of international standards at potential growth centres and to give boost to exports. The government has approved 11 apparel parks till now.

2.2.7.4 *Foreign Trade Policy*

The provisions of foreign trade policy, which may affect growth of textile industry favorably are –

- EOU's manufacturing textile/apparel are allowed to dispose off the leftover material up to 2% of import on payment of duty on transaction value.
- Import of second hand capital goods of any age has been allowed which would enable fresh capital investment in modernization of industry.
- For handicraft/handloom units dutyfree import of trimmings/embellishments raided to 5% of FOB value of exports, which would be exempt from countervailing duty.
- All exports oriented units (EOU) has been exempted from service tax in proportion of their exports.
- Under 'Target Plus', units with quantum growth in exports would be entitled to duty-free credit based on incremental exports.

2.2.7.5 *Union Budget 2005-06*

The various provisions in union budget 2005-06 which will provide boost to the industry and encourage fresh investment, upgradation of technology etc are –

- Reduction of customs duty on specific textile machinery from 20 percent to 10 percent
- Option of CENVAT exemption route is maintained for natural fibres including cotton.
- De-reservation of knitting industry from SSI category.
- Additional allocation of Rs. 435 crores for TUF Scheme.
- Reduction of excise duty of PFY from 24 to 16 percent.

2.3 SWOT ANALYSIS

The textile industry adds 14 percent to industrial production and 8 percent to Gross Domestic Product (GDP) of India. It provides direct employment to 38 million people (it is second largest employment provider, after agriculture). Indian textile industry is the world's 2nd largest producer of cotton yarn and 5th largest producer of synthetic fibre. Despite this strong presence, India's share in world trade in textile and apparel is a meagre 2.97 percent. The Indian apparel textile industry is one of the largest sources of foreign exchange flow into the country with the apparel exports accounting for almost 21 percent of the total exports of the country. The industry is vast with over 70,000 readymade apparel-manufacturing units and employs nearly three million people. Indian apparel export business has made great strides in the past few years and today many of the leading fashion labels, from all over the world, are known to source their products from India, but India still has only 2.8 percent of world apparel market and 4 percent of world textile market. A systematic analysis of textile and apparel industry indicate the following –

2.3.1 Strengths

There are number of factors which have led to the growth of Indian textile and apparel industry and can be termed as strength of industry.

(i) Raw material base

India has high self-sufficiency index for raw material particularly natural fibres. India's cotton crop is the third largest in the world. The Indian textile industry produces and handles nearly all types of fibers (both natural and synthetics). However the industry is pre-dominantly cotton based with 70 percent of the raw material being cotton. In recent years, India has emerged as the major producer of synthetics with large capacities being added. India produces world's 12 percent fibres following China (25 percent) and US (21 percent). India also has a well-developed woolen and silk industry.

(ii) Labour

Cheap labour and strong entrepreneurial skills have always been the backbone of Indian textile and apparel industry. It is the single largest employer in the industrial sector employing about 38 million people. If employments in allied sectors like ginning, agriculture, pressing, cotton trade, jute, etc. are added then the total employment is estimated at 93 million. Labour is a large component of the manufacturing costs in the textile sector. India has a competitive advantage due to low wage rates. India has one of the lowest labour costs in the world ($ 0.58 / hour). Table 2.31 indicate that labour cost in US is $ 14.24/hr. and $26.1/hr. in Japan while labour costs are $0.69/hr. and $0.37 respectively in China and Pakistan.

Table 2.31: Labour cost comparison

Country	US$ / hr
US	14.24
Japan	26.1
China	0.69
India	0.58
Sri Lanka	0.46
Bangladesh	0.43
Pakistan	0.37
Germany	8.3

Source: Compiled from Compendium of Textile Statistics

(iii) Flexibility

The small size of manufacturing which is pre-dominant in the apparel industry allows for greater flexibility to service smaller and specialized orders. Since the turnaround time for new product development is comparatively short in this Industry, the large base of small firms in India provide flexibility for producing small quantities and servicing to niche markets.

(iv) Rich heritage

The cultural diversity and rich heritage offers good inspirational base for designers. The fabric and the work of artisans from Karnataka, Tamilnadu, Rajasthan, North-east states etc. are well known in world textile and apparel trade.

(v) Domestic market

Natural demand drivers including rising income levels, increasing urbanization and growth of the purchasing population drive domestic demand. The variable demand drivers include latent demand in existing applications of textiles (good quality cloth at reasonable prices) and new applications of textiles. The extent of latent demand for good quality apparel can be gauged by the explosive growth of the nascent branded apparel segment, which has been growing at 15-20 percent in the last few years.

2.3.2 Weaknesses

There are a few area(s) due to which textile and apparel industry has not been able to perform well in world market. The major causes of its weak performance are:

(i) More dependence on cotton

More than 80 percent of the fiber consumed in India is cotton while in the world consumption market reverse trend is observed. This over dependence on cotton affects the textiles apparel trade in India. Due to over specialization in cotton, the bulk of international trade is missed out, synthetic products in India are expensive and fabric required for items like swimwear, ski-wear and industrial apparel is relatively unavailable, whereas world trade in synthetics is growing at the fastest rate, India has been unable to diversify its exports basket and remains heavily dependent on cotton. Table 2.32 shows that apparel export of cotton based apparel has increased while of non-cotton apparel has decreased in last decade.

Table 2.32: Fibre-wise split of India's Apparel Exports

Fibre	1994-95	1999-2000	2002-2003
Silk	1.9%	0.7%	0.6
Noncotton	29.3%	29.6%	13.4%
Cotton	68.8%	69.7%	86%

Source: Compiled from Compendium of Textile Statistics

(ii) Spinning sector

As discussed earlier, the spinning sector has excess capacity and has seen good growth but it lacks in modernisation. There is need for scraping of old spindles and introduction of new technology.

(iii) Weaving sector

India has relatively lesser number of shuttleless looms. The total number of shuttleless machines is currently estimated at around 20,000 while China has reached a level of approximately 100,000 shuttle less looms. The share of shuttleless machines in total weaving output is perhaps the lowest in India, in percentage terms. Looming robots are used in the textiles and apparel units in the western countries. Only 1.69 percent of total looms installed in India are shuttleless while in China the percentage of shuttleless looms is 15.29 percent of total looms. Table 2.33 reflects extent of modernization of weaving industry; Whereas Indian Industry is quite outdated in comparison to US, Pakistan, China, Brazil and Mexico.

Table 2.33: Shuttleless looms as a percentage of total installed looms

Country	Shuttleless looms (%) 2003
China	15.29
Indonesia	10.49
India	1.69
Brazil	29.54
Japan	16.52
Mexico	28.95
US	74.15
Pakistan	8.82

Source: Compiled from Compendium of Textile Statistics

(iv) Fabric processing

Processing stage add significant value in the entire textile chain, however, processing is the weakest link in the Indian textile value chain, adversely affecting its ability to compete in exports. India's export performance in terms of proportion of greige and processed fabric along with the UVR is shown (Table 2.34). The consumption pattern of world market indicates that 73 percent of consumption is of processed fabric and 27 percent of greige fabric whereas India's exports of fabric consist of only 34 percent processed fabric. This indicates an opportunity loss due to lack of adequate processing facility and technology.

Table 2.34: Proportion of greige to processed fabric

Country	Processed	Greige
India	34%	66%
World	73%	27%

Source : Textile Outlook International

The UVR of processed fabric is comparatively very low for India in comparison to average UVR for processed fabric in world market.

The UVR of processed fabric (more than 200 g/m) is 1.09 US$ for fabric exported from India while the average price realized in world market for similar specifications is US$ 2.01. There is comparatively higher gap in UVR of processed fabric in comparison to greige fabric being exported to world market. The underlying reason lies in technology gap existing in processing industry of India vis-à-vis. world (Table 2.35). The analysis indicates either India is not being able to cater to requirements of processed fabric and concentrating on greige fabric. Moreover the quality of processed fabric is rather inferior resulting in lower price realizations.

Table 2.35: UVR for (Greige/Processed) fabric

(Value in US$)

Weight Range	India	World
Greige, less than 100g/m	0.41	0.5
Greige, 100-200g/m	0.67	0.76
Greige,, More than 200g/m	0.73	0.96
Processed, less than 100g/m	1.51	1.84
Processed, 100-200g/m	1.20*	2.19
Processed, more than 200g/m	1.09	2.01

Source: Textile Outlook International.

(v) Garmenting

The garment industry is highly fragmented and is characterized with small and medium size units. This gives the industry flexibility to cater to smaller orders with variety but lacks in fulfilling huge orders for single quality.

(vi) Poor Infrastructure

High power costs (power cost is defined as number of units required to produce one kg. of yarn or one kg. of fabric) and long export lead times are eroding India's export competitiveness across the textile chain. Though India's power costs are higher than its competitors, the quality of power is significantly poorer.

(vii) Low labour productivity

Labour cost is lower in India but productivity levels are rather poor in comparison to other key exporting countries. The productivity levels for manufacturing various apparel items are far lower in India vis-à-vis its competitors. For example, on an average 9.1 pieces of Gent's shirts per machine are being produced in India while 20.9 Gent's shirts are being produced per machine in Hong Kong in a day. The poor productivity nullifies effect of comparative advantage in low wage cost.

(viii) Training

India has a labour intensive garment sector, with the advent of technology manpower requires appropriate training. An average of only 10 hours of training is given to the workers as against 70 hours of training given to the workers in China. There is poor concern and understanding of non-tariff issues like child labour, environment, and crèche availability for women employees etc.

(ix) Poor quality standards

Raw material quality, obsolete technology used for production, improper material handling techniques result in value loss of the goods. Many importers see India mainly as a source for low priced apparel for lower end of the market. This exerts a downward pressure on export prices hence UVR is low. Quality is not uniform in many cases hence there is a lack of confidence in India's products. Total quality Management is ensured in Japanese and American plants while it is still largely absent in the Indian industry.

(x) Distance of the potential markets

More than 80 percent of Indian exports is to US and EU. With the opening up of the world economies new challenges come up for the industry, the most important one being other emerging markets. There is a need for quick response, consequently, supplies in the two major consuming regions are shifting end product manufacturing to nearby low cost countries. Due to the distance of these markets from India and due to the increasing fashion risk these countries are shifting there sourcing base from India to neighboring developing countries even at a slightly higher price. In some cases, US is shifting its souring base to Mexico, Caribbean and Latin American countries. EU is looking towards east European & North African countries for apparel imports.

(xi) Lower average consumption in the domestic market

Per capita cloth consumption in India is almost 1/8th to 1/10th of those in the developed countries. This reduces the potential of the domestic market for apparel and textile manufacturers. This can be attributed to lack of good quality products at reasonable prices and lack of organised retail sector to create market "pull"

(xii) Lack of professionalism and integration of supply chain

Conflict and competition between small, medium and large players and amongst links of supply chain namely viz, cotton producers, spinner, weavers etc. The industry faces a shortage of professionals and the family business structure still pervades.

(xiii) Dependence on quota system

With the abolition of the quota system under the MFA, India is likely to loose its share in the world market to others who are better equipped for competition. Moreover, India's destination for exports is primarily quota market, which is going to have more effect of phasing out of quota. On the other hand countries i.e. China has developed good market for its product in non-quota countries too, resulting in possibility of gain and sustenance in those markets in longer period. Recent liberalisation with the government following a protectionist external policy, India is new to open market operations and was until protected from MNC's is now facing stiff competition. The market is set to witness more and more brands and product from US, EU and China threatening domestic market. The domestic market needs to be ready to face the challenges thereof.

(xiv) Very low investment in R&D

Firms in Southern US are reported to be researching the use of genetic engineering, cellular biology and tissue culture to grow coloured cotton. Practically no research is being done in the field of development of modified fibres like tencel, spandax, etc. India is still continuing with the slow methods of production, which results in lesser production and the products, do not meet the world standards in quality. Since in the past there was no need to change the product etc. but now there is need for innovation in terms of styling, packaging and product features.

(xv) Limited exploitation of economies of scale

Most of the players in the industry are small firms operating with relatively old technology. There is a shortage of large production lines to tap the basic product markets and to service huge orders on time. In spite of capital availability, hesitation of players to recruit large workforce (with the fear of union, idle manpower due to lack of orders) resulted in sub-optimal scale of operations. Large Indian units typically employ 1000-1500 workers; large units in China employ upto 25000 workers. India accounts for about 21 percent of the world's spindleage (second largest after China) and 58 percent of the world's loomage. The capacity utilisation in the spinning sector of the organised textile mill industry increased from 80 percent in 1990-91 to 86 percent in 1995-96 but it again decreased to the level of 83 percent during the year 2003-04, while the capacity utilisation in the weaving sector of the organised textile mill industry has remained less than 60 percent in last decade. The spindleage increased from 11 million in 1951 to over 37.03 million and rotors from 45 thousand in 1989 to 395 thousand in 2002; loomage however, declined from 1.78 lakh in 1990 to 88000 in 2004 in the organised sector. The percentage utilization of looms has reduced from 58 percent to 53 percent during 1990-2004.

Table 2.36: Capacity utilization in textile industry

Year	Installed spindlesNos (in million)	Percentage utilization	Installed loomsNos. (in thousand)	Percentage utilization
1980-81	21.23	77	208	78
1990-91	26.67	80	178	58
2000-01	37.91	85	123	47
2001-02	38.32	82	123	42
2002-03	39.03	80	119	41
2003-04	37.03	83	88	53
% Change 1981-2004	74.42	7.79	- 57.69	- 32.05

Source: Compiled from Compendium of Textile Statistics

As shown in Table 2.36 the number of spindles have increased by 74.42 percent between (1981-2004) while the percentage utilization has only increased by 7.79 percent in corresponding period. While number of installed looms has decreased by 57.69 percent and the percentage utilization (weaving) has reduced by 32.05 percent in corresponding period. This shows that although capacities are added in spinning but due to lack of modernization of the existing capacity in weaving, its capacity utilization is rather poor in weaving sector. Further to it, the weaving sector has not seen any additions in the capacity showing lack of interest in weaving sector.

Besides it, the other weaknesses include red-tapism leading to various problems like delay in shipment, loss of business and image hampering image of India as brand in exports besides effecting prospects of garment exports. In addition to it, Labour laws are very strict and rigid do not favour the manufacturing industry.

2.3.3 Opportunities

The external environment of world textile and apparel is expected to have growing market and there is an opportunity for India in textile and apparel industry due to;

(i) Growing industry

World textile trade would continue to grow at a rate of 3-4 percent to reach $200-210 billion by 2010 from $157 billion in 2002. Home textile and technical textile will grow faster than other categories.

(ii) Market access through bilateral negotiations

The trade is growing between regional trade blocs due to bilateral agreements between participating countries. India has no pacts with any country / trading block. Non-participation in these trade agreements is resulting in erosion of cost-competitiveness due to additional incidence of duty.

(iii) Opportunities in terms of new lucrative markets

Traditionally quota markets are India's key destinations but there are new and emerging markets i.e. South Africa, which are characterized by high growth, rate and has scope for market expansion for exporting country.

(iv) Integration of information technology

'Supply Chain Management' and 'Information Technology' has a crucial role in apparel manufacturing. Availability of EDI (electronic data interchange) makes communication fast, easy transparent and reduces duplication. Upcoming technologies for mass customization such as three dimensional non-contact body measurement and digital printing can give competitive advantage to the manufacturers. Global integration of supply system in a cost and time effective manner has become the need of the day. Inventory planning, sales forecasting, manufacturing strategy, distribution network and transportation management are some of the areas that will improve productivity and remove bottlenecks.

(v) MFA phase out

Post-2004, due to MFA phase-out India is expecting to have improved market access to major consuming markets in US and EU. The quantitative restrictions in terms of quotas are restricting entry of products of manufacturers without quota. The phase out of quotas shall be a boon to those exporters.

(vi) Opportunity in higher value items

The average UVRs of items from Hong Kong are 14-84 percent higher than UVR for similar products from India (Table 2.37). India also has the opportunity to increase its UVR's through moving up the value chain by producing value added products and/or by producing more value added technologically superior products.

Table 2.37: Comparison of UVR

Category/Country	Average UVR (US$)
Women's dresses	
Hong Kong	7.9
India	6.9
Women's blouses	
Hong Kong	7
India	3.8
Men's Shirts	
Hong Kong	6.3
India	4.7

Source: Compiled from CII Accenture "Textile Industry: Road to Growth" Report–2002.

2.3.4 Threats

The industry has good growth in past few years but a few factors pose a threat to prospects of Indian textile and apparel industry.

(i) Decreasing fashion cycle

There has been an increase in the number of seasons per year. This has resulted in shortening of the fashion cycle. Shortening of the fashion cycle has increased the fashion risk. This has caused US and European buyers to buy from the neighboring countries to hedge this fashion risk even at a higher cost

(ii) Formation of trading blocks

Formation of trading blocks like NAFTA, SAPTA, etc. has resulted in a change in the world trade scenario. Existence of bilateral agreements would result in significant disadvantage for Indian exports. NAFTA and EU are becoming self-sufficient blocs with growing trend of outward processing trade (OPT) resulting in increase in trade within these blocs (growth of 17 percent of trade from Eastern Europe to EU;

growth of 23 percent of trade from Latin America to North America). Developing countries in these trading blocks get a preferential treatment from the developed countries, which form the most lucrative markets for the Apparel exports. In addition to sourcing strategies, majority of other countries are also aligned to one or more trading blocs; India has no pact with any trading bloc/country. India's non-participation in these trade agreements is resulting in erosion of cost-competitiveness due to additional incidence of duty (Indian export have disadvantage of' 15-35 percent due to these restrictions). Pakistan has entered into a bilateral agreement with EU where for providing improved access to EU imports, Pakistan is getting concessional tariffs on its exports to EU.

(iii) Phasing out of quotas

The increasing competition from major exporting countries like china, Hong Kong, Korea, etc has opened the trade so the manufacturer, which provides the best deal, will flourish. India will have to open its protected domestic market for the foreign player's thus domestic players have to face a stiffer competition. India's proportion of exports in Top-10 quota countries is 3.2 percent as compared to 1.6 percent in Top-10 non-quota countries while China has 38 percent share of Top-10 non-quota countries and in 11.3 percent in non-quota countries. Quotas have distorted industry demand and supply dynamics, resulting in consumers in developed countries paying higher prices for apparel. This is also substantiated by India's price realisation for non-restrained items, which are substantially lower than realisation for restrained items. India's inability to gain market share in non-quota countries implies that quotas are currently protecting India's market share in apparel.

The strategies for post MFA period are to be designed with a perspective to capitalize on strength and utilize the opportunities to be available in quota free world. The strategy also needs to be formulated to overcome the weakness of textile and apparel industry and consolidate its position in world textile and apparel industry.

2.4 SUMMARY

India Apparel & Textile industry is one of the largest net foreign exchange earner for the country. Indian Textile industry is classified into several segments based on the type of establishment or by the fiber mix. The number of spinning mills have increased in last decade while the number of weaving mills is stagnant. Weaving sector is characterised with large presence of unorganised powerloom sector. The fabric from powerloom is lacking in uniformity in quality standards besides problem in delivering quantity production. Processing is one of the weak links in the textile supply chain of Indian industry. Indian apparel industry has witnessed vast growth in last decade. The industry consist of small scale units due to various government policies in past having provision for reservation of sector under SSI. However, with change in government policies and steps by industry fresh investments for large and vertically integrated units are seen in this sector. India is well positioned to take advantage of quota phase out given its unique strength which should give it long term competitive advantage.

Chapter 3

Apparel & Textile: Global Competitive Scenario

3.1 COMPETITIVENESS

Competitiveness is about productivity, which in turn is a function of factors related to cost of products, as well as those related to non-price factors such as delivery schedules, reliability of producers, and such intangible factors like image of the country/company and brand equity. Together, they define the competitiveness of a product to compete under conditions of free market.

The competitiveness of the textile and apparel industry in a country or region depends not only of the core competence of individual enterprises in the industry, but on the integration of the whole supply chain and relevant supporting industries, as well as internal and external business environments.

The core competitiveness of individual enter-prises is the fundamental elements for the sustainable competitiveness of an industry. Prahalad and Hamel (Hamel 1990) have pointed out that current competitiveness of a company derives from the price/performance of existing products. Future competitiveness derives from the ability to build, at low cost and more speedily than competitors.

The textile and apparel industry is set to witness a tough competition in post quota period. The existence and growth of industry shall depend upon various factors, the key being competitiveness in world trade. The various factors leading to contribute in competitiveness are discussed in this chapter.

Literature survey on Competitiveness of the Industry

In the changing world textile and apparel trade scenario, the transition from comparative advantage to competitive advantage has become essential. Porter (1990) believed that classical trade theories have been inadequate to explain trade patterns in today's global market in terms of how certain industries in certain nations succeed while others do not. The concept of competitive advantage has been developed to overcome this lacuna. In this context, the observations of Porter (1990) assume significance. "Competitive advantage that rests on factor costs is vulnerable to even lower factor costs somewhere else or governments willing to subsidize them. Today's low labour cost countries are being rapidly displaced by tomorrows. The lowest-cost source for a natural resource can shift overnight as new technology allows the exploitation of resources in places hitherto deemed impossible or uneconomical".

Porter (1990) examined why a nation achieves international success in a particular industry and concluded that four broad attributes of a nation shape the environment and these attributes promote or hinder the creation of competitive advantage. India need to move towards building these higher order advantages especially through favorable conditions of economic development specifically in industries which are favorably placed like garments, the present study focuses on sources of competitive advantage as well as conditions which will help the comparative advantages to expand and grow. Porter (1990) said, "Geographic concentration of firms in internationally successful industries occurs because the influence of the individual determinants in the diamond and their mutual reinforcement are heightened by close geographic proximity within a nation. An initial advantage in factors of production often provides the seed for an internationally competitive industry or a predecessor industry in the cluster". In the long run, firms succeed related to their competitors if they posses sustainable competitive advantage.

If we examine the four determinants, Cluster becomes very important for the success of an industry in a country. A cluster is formed when all related industries come up locally due to various factors. These clusters act as an umbrella for one particular industry and a firm; and help to accentuate the value chain of each other. In India, this value chain is missing. As far as demand conditions are concerned, it is observed that world-class domestic market produces a world-class industry and a world-class industry produces world-class competition (Prahalad, 1990). As far as India's garment industry is concerned the demand conditions within the country have not helped creation of a world-class market and this has blocked the development of a world-class garment industry. Newton (1991) observed, "The business of the clothing industry is to produce fashion and functional clothing. There is strong inter-relationship between the two areas and they strongly influence each other. Clothing is an important part of the fashion business in which design of the product is marketed as fashion. Fashion is the essential element of design, that is, dependent on timing and acceptance".

In the case of export of garments, the design quality, functional quality of the product and timeliness of fashions which mean indirectly delivery of the garments at the right time and related aspects all form the multi-dimensional constituents of quality. The sustainability of competitive advantage depends upon the 'source of advantages' and the hierarchy of 'competitive advantages' in terms of sustainability. Lower order advantages such as low labour cost or cheap raw material are relatively easy to imitate. Higher order advantages, for example, product differentiation or brand reputation or unique products are difficult to imitate. The second determinant of sustainability is the distinct sources of advantages a firm possesses. The second determinant of sustainability is the distinct sources of advantages a firm possesses. The third and most important competitive advantage is constant improvement and upgrading. Sustainability is achieved when the advantage resists erosion by competitor behavior (Porter, 1985). In other words, the skills and resources underlying a business's competitive advantage must resist duplication by other firms. Ghemawat (1986) held that the sustainable advantages fall into three categories: (a) Size of the target market, (b) superior access to resources or customers and (c) Restrictions on competitor options. The advantages are non-exclusive. Competitive advantage grows fundamentally out of the value a firm creates to its buyers and creating value for buyers that exceeds the costs of doing so is the goal of any generic competitive strategy. Value is what buyers are willing to pay for what a firm provides them and superior value stems from offering lower prices than competition for equivalent benefits or providing unique benefits that more than offset a higher price.

The core competencies of individual enterprises are the fundamental elements for the sustainable competitiveness of an industry. Prahalad and Hamel (Hamel 1994) have pointed out that the current competitiveness of a company derives from the price/performance of existing products. Future competitiveness derives from the ability to build, at low cost and more speedily than competitors. (Li Yi and Edward Newton, 2000). At the level of core competence, the goal is to build world leadership in the design and development of a particular class of product functionality (Hamel 1994).

In terms of regional competitiveness, Porter (1998) points out that the enduring competitive advantages in global economy lie increasingly in local things, which distant rivals cannot match, including knowledge, relationships and motivation. The modern economic map of the world is considered as dominated by clusters, defined as geographic concentrations of interconnected companies and institutions that achieve unusual competitive success in a particular field. Porter and other researchers further argue that productivity, not exports or natural resources, determines the competitiveness and prosperity of any state or nation. The birth and growth of clusters depend closely on their productivity growth, which is largely determined by the local competition and innovation. This highlights the importance of industrial supply chain integration, support of relevant industries and general business environment.

Export competitiveness can be measured using the market share, relative price ratio, relative factor productivity ratio & quota utilization (Prasad Ashok Chandra, 1997). For the purpose of this study, industry has been defined as group of firms manufacturing products (Fabrics, Clothing) that directly or indirectly compete. It is implied that no nation can be competitive in manufacturing all goods and services. Since, it is the firms who compete in international market (Porter, 1998), the entire framework of competitiveness would revolve around the study at the firm. "Industrial success was founded on behavior of the firms, not on decision of governments" (Kay, 1996).

Storper (1992) has pointed out that in growing world trade, export specialization based on specific products becomes increasingly important. Export specialization is largely due to product-based "absolute" technological advantages, renewed through learning in a variety of dynamic economies. Such export-oriented absolute advantage industries tend to be organized into production and distribution networks combining the advantages of specialization and flexibility, called "technology districts". Features of such technology districts are: (1) Trade specialization is achieved by obtaining absolute technological scarcity of the products; (2) The technological scarcity is gained through technological dynamism in products through continuous learning; (3) Production networked are organized on the basis of elaborate shifting division of labour between firms or between units of a single firm, to achieve technological dynamic flexibility; (4) Key collections of physical, capital, labour and information resources for the production network are highly geographically concentrated in one or a few sub-national regions of the host countries; (5) The technological learning rests on the conventions of the regional production system which guide the mobilization and maintenance or resources in mutual engagement between firms. The conventions are rooted in local political, cultural and other non-economic forces, which determine the quality of the technology districts.

3.2 COMPETITIVE POSITION OF INDIA

As discussed earlier, the percentage share of textile and apparel in world trade has increased from 2.14 percent in 1980 to 5.1 percent in 2004. The growth is particularly high from year 1980-1995 while it is almost static from 1995 onwards till 2004. Out of this, the share of textile trade has increased from 1.23 percent to 2.2 percent while the share of apparel has increased from 0.9 percent to 2.9 percent showing a sharp increase in apparel trade in last two decades (Exhibit 3.1). The increase in share of textile and apparel in world trade can be attributed to

Exhibit 3.1: Apparel and Textile in world trade

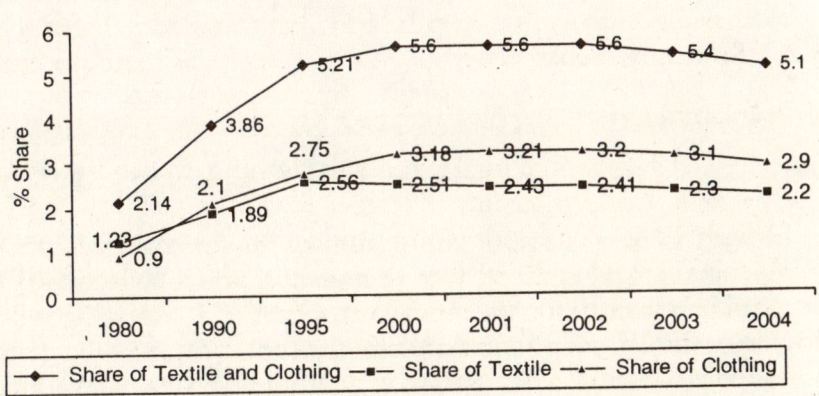

Source: Compiled from WTO

increased share of apparel trade in world. The growth in apparel trade is far ahead than in textiles in last two decades. This shows shift of world textile and apparel from textile to apparel.

India's share of textile trade in world has increased from 2.34 percent (1990) to 4.0 percent (2004). While apparel trade has increased from 2.09 percent to 2.80 percent. It shows only a marginal increase in India's apparel trade vis-à-vis world trade in apparel (Exhibit 3.2). The trend indicates world trade is towards apparel; India's trade still has more concentration of textiles in comparison to apparel.

The above indicates that although the world trade is growing faster in apparel while India continues to have comparatively better performance in textile trade. Since trade in apparel reflects value addition and integration of value chain i.e. textile and apparel. The Indian industry seems to lack in capitalizing on its strength in textiles for improving the performance in apparel industry.

Exhibit 3.2: India's share in world trade

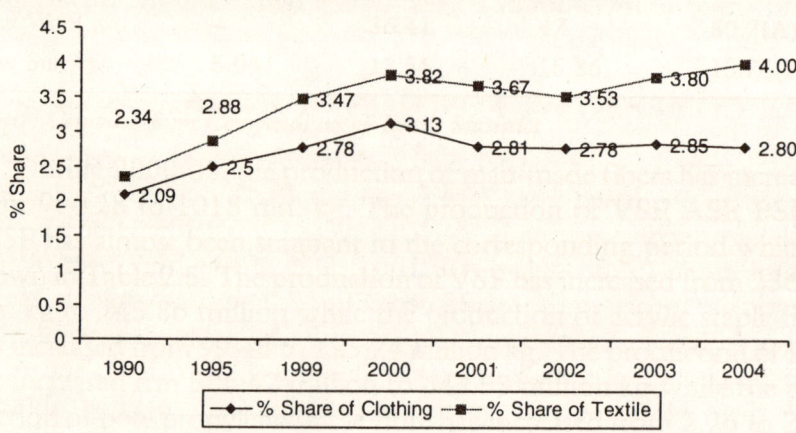

Source: Compiled from WTO

3.2.1 Spinning sector

(I) Fibre Mix

India has 3.13 percent of total textile and clothing export of world and has 7th rank. China is leading in textiles & clothing exports in world. India with 12.49 percent of world production is at 3rd rank in world while china is having 1st rank in raw cotton, silk, cellulosic and synthetic fiber yarn production. India has 2nd rank in silk and cellulosic fiber in world. It ranks first in jute production(60.6 percent) while in wool India has only 2.25 percent of world production and is having 8th rank. Australia is having 1st rank in wool production.) China is leading in spun yarn, cotton fabric followed by India. India contributes 11.03 percent of total world production of cotton yarn and India has 14.69 percent of total world production of cotton fabrics.

Table 3.1: India's Position in World Textile Economy

Parameter	Unit	World	India	India as % of world	India's rank in world	Country with 1st rank
Fiber/Yarn						
Raw cotton (2001 –02)	Mn no	21505	2686	12.49	3	China
Cellulosic fiber yarn (2002)	Mn no	2118	286	13.50	2	China
Synthetic fiber yarn (2002)	Mn no	27959	1759	6.29	5	China
Wool (greasy) 2001 – 02	Mn no	2262	51	2.25	8	Australia
Silk 2000	Mn no	81	15	18.52	2	China
Jute (2001-02)	Mn no	3119	1890	60.60	1	India
Total	Mn no	57044	6687	11.72		
Spun yarn – 2001						
Cotton yarn	Mn .kg	20087	2216	11.03	2	China
Fabrics – 2001						
Cotton fabrics	Mn .kg	11985	1761	14.69	2	China
Total fibers – 2000	Kg.	8.5	6.2	–	–	–
Total textiles & clothing exports	Bn US$	364.42	11.41	3.13	7	China

Source: Compiled from Compendium of Textile Statistics

Table 3.2 indicates that internationally there is more demand of textile and apparel with man-made fibre composition (58 percent of

Table 3.2: Fibre mix

	Cotton (2000)	MMF (2000)	Cotton (2004)	MMF (2004)
International demand	42 %	58 %	42.99%	57.01%
India – Domestic consumption	46 %	54 %	57.35%	42.65%
India – Exports	86 %	14 %	79%	21%
India – Production	76.3%	23.6%	76.9	23.1%

Source: Compiled from Compendium of Textile Statistics

world consumption) while India exports primarily cotton (79 percent
of total exports). The composition of pattern of production of textiles
in India indicates more focus on cotton whilst the demand in Indian as
well in world is more towards blended or 100 percent MMF.

(II) Spinning Capacity

India and China have one of the largest capacity for spinning yarn in
terms of total percentage of world capacity followed by Pakistan,
Indonesia, Turkey and Brazil having significantly lesser capacity of
spinning (installed spindles). The percentage of distribution of spindles
is shown in Table 3.3. India ranks Ist in number of spindles in cotton
system with 24.29 percent of world capacity while it ranks 2nd in terms
of overall spinning capacity with 22.7 percent of world capacity preceded
by China. India has 5th rank (5.55 percent of world capacity) in rotor
spinning. China followed by Pakistan, Indonesia & Russia are ahead of
India in this category.

Table 3.3: Comparison of Installed Spinning Capacity

Items /Country	India	China	Pakistan	Indonesia	World total
Spindles (mn. No)					
a) Cotton system	38.64	36.24	9.25	8.61	159.05
b) Woolen system	1.04	3.69	0.04	0.11	15.76
c) Cotton and woolen system	39.68	39.93	9.29	8.72	174.81
Ranking	2	1	3	4	–
Contribution to the world (%)					
a) Cotton system	24.29	22.79	5.82	5.41	100.00
b) Woolen system	6.60	23.41	0.25	0.70	100.00
c) Cotton and woolen system	22.70	22.84	5.31	4.99	100.00

Items /Country	India	Russia	China	US	World total
Rotors/ ('000 No)	464	2205	942	712	8293
Ranking	5	1	2	3	–
Contribution to the world %	5.60	26.59	11.36	8.60	100.00

Source: Compiled from Compendium of Textile Statistics

The large spinning capacity provides strength to India, if being utilized effectively to produce value added fabric and apparel as per end customers' requirement. The export of yarn provides foreign exchange to yarn manufacturer but does not help textile and apparel industry. This coupled with need to concentrate on finer counts make India competitive but it requires modernization of spinning industry too.

3.2.2 Weaving sector

Table 3.4 indicates that India (33.37 percent) has world's largest weaving capacity as a percentage of total world capacity. The other countries including China (19.82 percent of world capacity) and Pakistan (5.86 percent of world capacity), Indonesia (5.50 percent of world capacity) are far behind in terms of weaving capacity. India ranks 1st in 61 percent of capacity of weaving while in terms of shuttlelesss looms India has 9th rank in world (2.83 percent of world capacity). China is having 1st rank in this category with 17.7 percent of world capacity of Suttleless looms.

Table 3.4: Installed Comparison Capacity of Weaving

Items /Country	India	China	Pakistan	Indonesia	World
Shuttle less looms	21468	134173	18507	27356	758485
Shuttle looms	1803755	808796	260100	234520	3998724
Total	1825223	942969	2788607	261876	4557209
Ranking	1	2	3	4	–
Contribution to the world %	38.37	19.82	5.86	5.50	100.00
Ratio of shuttles looms to total loom age	1.18	14.23	6.64	10.45	15.94
	India	Bangladesh	Pakistan	Nepal	World
Handlooms	3900000	500000	80000	70000	4600000
Ranking	1	2	3	4	–
Contribution to the world	84.78	10.87	1.74	1.52	100.00

Source: Complied from Compendium of Textile Statistics

India contributes 84.78 percent of total world capacity of handloom and is followed by China (10.87 percent of world capacity).

India has only 1.17 percent of its total looms with shuttle less technology indicating lack of modern technology in weaving sector. This coupled with (Table 3.4) showing weaving capacity in competing countries indicate that although large weaving capacity is existing in India but it consists of old and outdated technology looms resulting in lesser production which is reflected with low utilization of weaving capacity. China has only 14.2 percent of the total world capacity but looms with shuttle less technology is relatively high. Indian weaving industry is in need for modernization and discontinuation of old looms but as discussed earlier, the number of weaving mills is stagnant and fresh investments are rather negligible.

3.2.3 Processing Sector

The average UVR of fabric export from Indian mills is 60 percent of world average price while powerloom fabric gets only 30 percent of average world prices. This can be attributed upto larger extent to more exports of greige fabric rather than processed one. India has gained strength and focused more on greige fabric due to lack of modern processing facilities. India's export of fabric primarily consists of greige fabric (66 percent of total) while World trade has 73 percent of consumption of finished fabrics (Exhibit 3.3). As discussed earlier, India is not competing in trade of processed fabric due to lack of modern technology while earnings are rather poor in greige fabric trade. The Indian processing industry is in need of upgradtion of processing technology and concentration on trade of value added processed fabric rather than greige fabric to be competitive in world textile and apparel trade.

Exhibit 3.3: Composition of world textile trade (greige/processed)

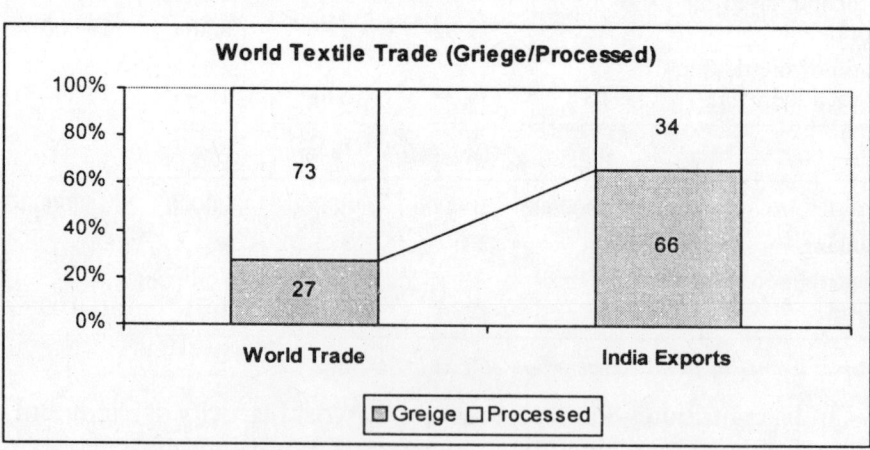

Source: Compiled from FICCI – KSA 2000

Competitive Analysis of Indian Textile Industry

Table 3.5 summarizes India's competitive performance in various segments of textile and apparel industry. Only 36 percent of India's spinning capacity is modernized while 90 percent of capacity of Italy and 50 percent of capacity in US is latest in technology. In weaving, India has only 20000 shuttleless looms while China has 90000 of shuttleless looms indicating high degree of modernization in weaving sector in China. India's processing technology is rather semi automatic while competitive countries have advanced and mechanized processing. Investment in research and development is only around 0.2 percent of total sale value in India while it is around 1.5-2 percent in competing countries. In China the number of machines per company is much higher (600 nos.) than India (120 nos.), which indicate that Indian industry is having lesser technological involvement in comparison to competing countries. In addition to it, the percentage of power machines is 60 percent of total capacity while in competing countries the extent

Table 3.5: Competitive analysis of Indian textile industry

Parameters	India	Competition
New spinning Capacity	36% of total	90% of total (Italy) 50% of total (US)
Shuttle-less looms	20,000	30,000 (Korea) 90,000 (China)
Processing technology	Manual/Semi-automatic	Advanced & Mechanised
Investment in R & D	0.2% of sale	1.5-2.0 % of Sales
Aggregation of capacity (machines / company)	120	600 (China) 260 (Korea)
Power machines – proportion	60%	98%
Yield of cotton (Kg. per hectare)	324	609 (Pakistan) 998 (China) 634 (World average)
Contamination in cotton	30-50%	15-18 % (World average)
Value added per worker (productivity)	1.91	2.53 (China)
Testing of materials and product	Adhoc & Client oriented	Integral part of system
ISO 9000 QMS certified units	462	3673
ISO 14000 EMS certified units	12	100
In-plant defect rates	3-5 %	< 1 % Canada 1-2 % China

Source: Compiled from "Background Paper", CII: India Textile Summit 2003

of power machines is around 98 percent The analysis indicates need for technological up gradation in all segments of the textile industry. In raw material (cotton) the yield of cotton is 324 per hectare in India while the yield in Pakistan is 609 kg. per hectare and in China 998 kg.per hectare. India's yield is much lower than world average of 634 kg. per hectare. Indian cotton is having 30-50 percent contamination while the world average is 15-18 percent, which indicates that Indian cotton has poor yield and high level of contamination. The gap in number of units which are ISO certified in India and competing countries is too wide which indicate poor quality adherence norms in India which is also reflecting in higher in plant defect rate in India.

3.2.4 Garment Sector

India's garment industry is characterised with fragmented structure having large number of units capable of producing small order sizes. However, there are many a units with substantial investments and fully integrated structure to cater large order (run) size. As per a study (Kathuria and Bhardwaj, 1998); the number of machines (average) installed (Table 3.6) by export firms indicates low penetration of technology in Indian industry. India has lowest figure of average number of machines in most of the stages of manufacturing of apparel. This coupled with low investment in technology (Table 3.7) shows India's poor preparedness in terms of technology in garment manufacturing. It is reflected that Hong Kong, South Korea and Taiwan have high investments per machine. In addition to it, Hong Kong, China and Thailand are having much more number of the machines in manufacturing of apparel than India.

Table 3.6: Machines installed by apparel export firms (average nos.)

Countries	Precutting machines	Cutting machines	Sewing machines	Special machines	Processing machines
S. Korea	2.9	12.3	134.3	77.5	31
Taiwan	2.6	7.5	185.1	49.5	12.8
Hong Kong	2.3	13.2	455.4	112.7	27.9
China	2.3	13.2	450.5	104.8	34.4
Thailand	2	12.8	460.8	72.4	21.9
India	0	2.3	103.7	8.6	4.6

Source: Compiled from ICRIER report, 2002; Kathuria and Bhardwaj, 1998

Table 3.7 reflects average number of machine with apparel firms in India as 119.28 while in china it is 605.15 and Hong Kong has 698.12

machines per firm. India has high number (37.26) of manual machines per firm in comparison to competing countries. The average number of power machines is 75.39 per unit in India while in China it is 603.65 and in Hong Kong the number of power machines per unit is 688.76 indicating poor level of technology in India. The investment per machine in India is one of the lowest in comparison to other competing countries in apparel trade.

Table 3.7: Machinery and investment by apparel export firms

Countries	Total machines (no's)	Manual machines (no's)	Power machines (no's)	Investment ('000$)	Inv. ('000$) per machine
S. Korea	258.08	6.14	240.33	722.19	2.79
Taiwan	264.62	0.15	264.46	579.21	2.18
Hong Kong	698.12	4.35	688.76	2456.64	3.51
China	605.15	1.5	603.65	9438.46	1.5
Thailand	572.32	0	562.32	722.25	1.26
India	119.28	37.26	75.39	29.76	0.25

Source: Compiled from ICRIER report, 2002; Kathuria and Bhardwaj, 1998

3.2.4.1 Comparison of Productivity

A comparison of productivity ratings (Table 3.8) of apparel

Table 3.8: Productivity rating of apparel manufacturing

Country	Productivity rating (%)	Country	Productivity rating (%)
Germany	100	Mauritius	70
France	100	Romania	70
US	100	Tunisia	70
UK	100	Malaysia	65
Italy	95	Thailand	65
Spain	90	Sri Lanka	65
HK	90	Vietnam	65
Taiwan	85	China	60
Korea	85	Pakistan	50
Greece	80	Indonesia	50
Poland	80	Cambodia	50
Turkey	75	India	50
Lithuania	75	Madagascar	45
Mexico	70	Bangladesh	40

Source: Compiled from 5th Cost Comparison study of KSA, 1999

manufacturing in different countries reflect that India has got a rating of 50 percent in productivity in comparison to Germany, France, US, UK (rating of 100 each).

The comparison of labour cost in Table 3.9 indicate that cost in India is 4 percent of labor cost in US. The labour cost is around or lesser than 5 percent of cost in US in China, India, Sri Lanka, Pakistan & Indonesia. Japan is having almost double of the cost than US making it most expensive in terms of Labour cost.

Table 3.9: Labour cost comparisons

Country	Total cost in $US	Ratio to US Cost %
US	14.24	100
Japan	26.1	183
New Zealand	7.28	51
Taiwan	7.23	51
Hong Kong	6.1	43
South Korea	5.32	37
Thailand	1.18	8
Malaysia	1.13	8
PR of China	0.69	5
India	0.58	4
Sri Lanka	0.46	3
Pakistan	0.37	3
Indonesia	0.32	2

Source: Compiled from UN COMTRADE. 2004

The performance of Indian garment industry in terms of productivity is rather poor in comparison to other competing industries (Table 3.10). Hong Kong is having productivity, which is around 100 percent more than the productivity in India for most of the categories in the apparel. The other countries with high productivity are Taiwan, Korea and Thailand. In ladies blouses category, Hong Kong manufactures on an average 20.6 pieces per machines per day followed by Taiwan (18.9), Thailand (17.0) while India produces only 10.2 pieces of ladies blouses per machine per day. The reason for productivity can be linked with Table 3.7 indicating high investment per machine and more number of machines per apparel firm.

Table 3.10: Productivity of apparel firms (pieces/machine/day)

Countries	Ladies blouses	Gents shirts	Ladies dresses	Ladies skirts	Trousers
Hong Kong	20.6	20.9	20.2	19.3	19.3
Taiwan	18.9	18.2	12.4	16.6	16.1
Thailand	17.0	19.8	12.2	20.5	13.1
S. Korea	14.6	17.4	8.8	17.5	15.6
China	10.9	14.0	7.8	13.0	6.7
India (CII study)	10.2	9.1	6.3	9.6	6.8
India (ICRIER)	10.18	9.12	6.25	9.62	6.84
India (NIFT)	10.3	9.4	5.7	9.3	6.8

Source: Compiled from CII Accenture "Textile Industry: Road to Growth" Report-2002, ICRIER- 1993, NIFT - 2000

3.2.4.2 Composition (Woven/Knitted)

An analysis of composition of world trade (woven vs. knitted) in Table 3.11 indicates that 60 percent of world trade is of woven apparel while around 40 percent is knitted apparel. One of the largest destination imports market i.e. US imports 40 percent (woven) and 60 percent (knitted) garment; while EU has 55 percent of trade in woven and 45 percent in knitted. It is evident that the trade in knitted category is increasing while India's exports consist of 60 percent of woven apparel and 40 percent of knitted apparel. As discussed earlier, US is the

Table 3.11: Composition of world apparel trade (% value)

	1999		2000		2001	
	Woven	Knitted	Woven	Knitted	Woven	Knitted
World (Import)	62	38	60	40	60	40
US (Import)	42	58	42	58	40	60
Germany Import)	58	42	56	44	58	42
Japan (Import)	56	44	56	44	58	42
France (Import)	58	42	55	45	56	44
EU (Import)	60	40	58	42	55	45
India (Export)	65	35	64	36	60	40
China (Export)	70	30	72	28	71	29

Source: Compiled from Handbook of Export Statistics, AEPC

single largest market of apparel. This coupled with analysis of Table 3.11 reflect high demand of knitted garment in US and indicate more potential of knitted garment market in US.

The composition of India's trade (woven vs. knitted) is shown in Table 3.12. It reflects that India's export in year 2001-02 consisted of 40 percent by value (knitted) and 60 percent by value (woven) apparel. However, in terms of volume knitted garment consists of 57 percent of total exports implying low UVR for knitted apparel. It can be interpreted that growth in exports of knitted garment 33.14 percent in last six years (volume wise) while growth in value terms is 19.71 percent in similar period. The export of woven garment from India is around is 60 percent in total value while 43 percent in volume terms indicating higher realization from export of woven apparel.

Table 3.12: Composition of India's apparel trade (% value)

Year	Knitted apparel		Woven apparel	
	Volume (mn pcs.)	Value (US bn $)	Volume (mn pcs.)	Value (US bn $)
1996-97	540.2(46%)	1.46(31%)	641.8(54%)	3.24(69%)
1997-98	632.4(49%)	1.59(33%)	664.0(51%)	3.29(67%)
1998-1999	682.0(51%)	1.62(32%)	669.0(49%)	3.54(68%)
1999-2000	780.1(54%)	1.95 (35%)	658.6(46%)	3.55(65%)
2000-2001	765.9(55%)	1.91(36%)	625.5(45%)	3.38(64%)
2001-2002	719.6(57%)	1.74 (40%)	547.2(43%)	2.66(60%)
% Change 2001-02/1996-97	33.14	19.71	-14.66	-17.9

Source: Complied from Handbook of Export Statistics, AEPC

3.3 COMPETITIVE POSITION OF INDIAN EXPORTS

3.3.1 US Market

China followed by Mexico are leading suppliers of MFA textile & apparel to US (Table 3.13) followed by Hong Kong, India & Canada, South Korea etc. India has 4th rank in supply to US in value tems and 6th rank in volume imports from US. The average price (US$/ piece) is 1.93 for India which is close to overall average price of imports to US (US$ 1.89). The average price of supply from China is $1.25 while the imports from Hong Kong have average price $4.59.

Table 3.13: Imports of MFA apparel and textiles

	1990	2000	2003
China			
Value (US$ mn)	3127	6527	14555
Volume ('000 sme)	1682	2218	11664
Price (US$/piece)	1.86	2.94	1.25
Mexico			
Value (US$ mn)	647	9693	7771
Volume ('000 sme)	432	4746	1081
Price (US$/piece)	1.50	2.04	1.90
Hong Kong			
Value (US$ mn)	3686	4707	3958
Volume ('000 sme)	957	1123	862
Price (US$/piece)	3.85	4.19	4.59
Canada			
Value (US$ mn)	417	3350	3062
Volume ('000 sme)	640	3204	3264
Price (US$/piece)	0.65	1.05	0.94
India			
Value (US$ mn)	743	2741	3631
Volume ('000 sme)	377	1248	1914
Price (US$/piece)	1.97	2.20	1.90
All US imports			
Value (US$ mn)	26749	71692	82844
Volume ('000 sme)	12144	32864	46628
Price US$/piece	2.20	2.18	1.78

Source: Complied from Office of Textile & Apparel, U.S. Dept. of Commerce

US being single largest market of apparel (41.6 percent of total world apparel trade) is most attractive destination along with EU (49.69 percent), which as a block of the countries is largest market in world apparel trade. Table 3.14 indicates that India is having V rank in US market preceded by China, Mexico, Hong Kong and Canada. India is having I[st] rank in Ladies dresses (cotton) and W&G shirts (cotton) category. Besides it, India has II rank in women's suit and III rank in Ladies blouses, M&B shirts (cotton) and women's overcoat category and IV rank in Ladies skirts (MMF) and M&B cotton knitted shirt category. However in W & G Trousers (cotton), M & B Trousers (cotton) and M & B cotton coats import from US, India is not having substantial share in US imports.

Table 3.14: Suppliers of apparel to US

Country/Product Category	Rankings				
	I China	II Mexico	III HK	IV Canada	V India
Ladies Blouse	Turkey	Romania	India	HK	Poland
M & B Shirt (Cotton)	HK	Bangladesh	India	Indonesia	Philippines
Ladies Dress (Cotton)	India	Philippines	HK	China	Sri Lanka
W & G Shirt (Cotton)	India	HK	Bangla-desh	Indonesia	China
Ladies Skirt (MMF)	China	Guetemala	Canada	South Korea	India
Ladies Skirt (Cotton)	HK	Mexico	India	Cambodia	Philippines
W & G Trousers (Cotton)	Mexico	HK	China	Philippines	Turkey
M & B Trousers (Cotton)	Mexico	Dominican Republic	HK	Guatemala	Bangladesh
W & G Cotton Knit Shirt	Mexico	Guatemala	Honduras	HK	Turkey
M & B Cotton Knit Shirt	Mexico	Honduras	Pakistan	El Salvador	India
Women's Suit	China	India	Turkey	Romania	Poland
Women's Overcoat (Woolen)	Italy	Dominican Republic	Philippines	Guatemala	HK
Women's Overcoat (Cotton)	China	HK	India	Philippines	South Korea
M & B Cotton Coat	China	HK	Mexico	Pakistan	Sri Lanka

Source: Compiled from Office of Textile & Apparel, U.S. Dept. of Commerce

The detailed analysis of competitive position of India in US imports for select apparel categories is as follows:

i) Man-made fibre skirts (Category 642)

India ranks IInd in value terms (7.4 percent share) preceded by China and followed by Canada & Guatemala and IIIrd in volume (7.7 percent share) of import from US preceded by Guatemala, Canada, Mexico. Average price of import from India is US$ 5.33 while average price of US imports is US$ 6.18 per piece. It is significant here that the US imports has increased in this category but import from India has decreased (Table 3.15). China is 1st in value and 8th in volume in this category with average price of $9.42 reflecting higher end of market catered by China in this category. Average price of imports from Canada ($11.24) and from Guatemala ($4.09) are in this category also reflect their respective target markets as both of them share 1st rank in volume imports from US for this category.

Table 3.15: Imports of Man-made fibre skirts

	1990	2000	2004
India			
Value (US$ '000)	22936	53395	38248
Volume ('000 sme)	4093	12497	7941
Price (US$/piece)	6.96	5.33	5.37
All US imports			
Value (US$ '000)	290936	639941	564475
Volume ('000 sme)	45576	128661	121457
Price (US$/piece)	7.93	6.18	5.95

Source: Complied from Office of Textile & Apparel, U.S. Dept. of Commerce

ii) Cotton skirts (category 342)

India ranks Ist in value followed by Cambodia, Hong Kong and Sri Lanka in US imports of cotton skirt. The trade (import from India) is IInd in in volume terms preceded by Combodia. The average prices have come down in last one decade from US$ 6.01 to US$ 4.89 and average earnings to India are lesser than average price of imports from US i.e. US$ 5.48 (Table 3.16). The average price of imports from Cambodia (Ist in volume & IInd in value in imports from US) is $4.83 while average price of imports from Hong Kong are $8.85. China has 10th rank in imports value of this category.

Table 3.16: Imports of cotton skirts

	1990	2000	2004
India			
Value (US$ '000)	29773	27284	84422
Volume ('000 sme)	6148	7175	20307
Price (US$/piece)	4.72	4.72	4.89
All US imports			
Value (US$ '000)	289584	425565	818368
Volume ('000 sme)	51957	92532	187629
Price (US$/piece)	6.92	5.71	5.48

Source: Complied from Office of Textile & Apparel, U.S. Dept. of Commerce

iii) Women's and girls' cotton non-Knit (woven) shirts (category 341)

India ranks I in import of women's & girl's cotton non-knit shirts with average price US$ 4.83. In this category, the average earnings are lower than average price of import (US$ 5.26) for this category. Table 3.17 shows that the India's trade in this category has increased by around 8 percent in value as well as volume terms. Hongkong, Bangladesh, Indonesia &Sri Lanka are other leading countries in this category in value. The average price of imports from China is $9.81 while its $3.68 from Macau.

**Table 3.17: Imports of Women's and Girls' cotton
non-knit (woven) shirts**

	1990	2000	2004
India			
Value (US$ '000)	139249	300132	326446
Volume ('000 sme)	33927	54969	761441
Price (US$/piece)	4.14	5.51	4.83
All US imports			
Value (US$ '000)	612639	1249260	1431784
Volume ('000 sme)	127794	226111	268411
Price (US$/piece)	4.83	5.57	5.26

Source: Complied from Office of Textile & Apparel, U.S. Dept. of Commerce

iv) Men's and boys' cotton non-knit (woven) shirts (category 340)

India ranks first in import of this category with average price US$ 4.55. In last one decade the trade for India has increased in volume terms by 6.5 percent but in value terms it has decreased marginally. At the same time the average price has reduced by 6.2 percent in last one

**Table 3.18: Imports of Men's and Boys' cotton
non-knit (woven) shirts**

	1990	2000	2004
India			
Value (US$ '000)	70513	188379	199418
Volume ('000 sme)	22489	45494	43815
Price (US$/piece)	5.25	6.94	4.55
All US imports			
Value (US$ '000)	1039426	2422956	2366032
Volume ('000 sme)	299359	606796	607010
Price (US$/piece)	5.82	6.69	6.52

Source: Complied from Office of Textile & Apparel, U.S. Dept. of Commerce

decade. In this category the average price of import in is US$ 6.52 (Table 3.18). Hong Kong followed by Bangladesh, Indonesia, Philippines and Sri Lanka are other leading countries in this category.

The average price of imports is $ 8.82 from Hongkong $ 4.17 from Bangladesh and $ 6.12 for imports from Indonesia, Philipenes.

v) Men's and boy's cotton knit shirts (category 29)

India ranks 6th in value terms and 8th in volume terms in imports from US of this category. The trade has increased by approx. 8 percent for India. The prices have come down marginally to US$ 5.34 per piece. Mexico, Honduras, Pakistan and El Salvador are other countries with substantial share in this category.

Table 3.19: Imports of Men's and boy's cotton knit shirts

	1990	2000	2004
India			
Value (US$ '000)	2695	243888	291262
Volume ('000 sme)	331	19368	27239
Price (US$/piece)	4.07	6.30	5.34
All US imports			
Value (US$' 000)	888289	4719216	5182366
Volume ('000 sme)	102451	717513	931938
Price US$/piece	8.67	6.59	5.56

Source: Complied from Office of Textile & Apparel, U.S. Dept. of Commerce

Honduras followed by Mexico and Pakistan & El Salvador, Vietnam are leading countries for this category imports from US. The average price of imports $2.13(Mexico), $1.53(Honduras) are much lower than prices from India. China has more than $7/pc but is at 10^{th} rank in value and 20^{th} in volume. Average import price is US$ 5.56 (Table 3.19).

vi) Cotton dresses (category 336)

In category 336, India ranks first (Table 3.20) in imports from US market. The trade has increased for India while the prices have reduced to US$ 5.87 (2004) from US$ 7.98 (1990). The average prices have reduced from $8.53 to $5.01 for US imports. The other competitors for India in this category include Philippines (2nd rank in imports of this category), China ($11.02) and 3rd in value import followed by Hong Kong, Sri Lanka, Pakistan has $1.79 and is 3rd in volume imports for this category.

Table 3.20: Imports of cotton dresses

	1990	2000	2004
India			
Value (US$ '000)	24170	46234	52998
Volume ('000 sme)	9562	17377	35924
Price (US$/piece)	7.98	8.40	5.87
All US imports			
Value (US$' 000)	278414	432454	355795
Volume ('000 sme)	103101	220417	225844
Price US$/piece	8.53	6.20	5.01

Source: Compiled from Office of Textile & Apparel, U.S. Dept. of Commerce

vii) Women's and girls' cotton coats (category 335)

India ranks second in volume terms and third in value terms (Table 3.21) with an increase of 8.4 percent trade in last five years. The average price of import has reduced substantially from US$ 14.72 to US$ 9.01 while earnings for India have increased from US$ 5.59 to US$ 9.20 during 1990-2004. China, Hong kong followed by South Korea, Thailand are major suppliers besides India in this category. The average price of imports from Italy ($55.43), China ($17.40) is much higher than import price from India.

Table 3.21: Imports of Women's and Girls' cotton coats

	1990	2000	2004
India			
Value (US$ '000)	7940	26287	68585
Volume ('000 sme)	4083	10315	21415
Price (US$/piece)	5.59	7.33	9.20
All US imports			
Value (US$' 000)	364808	372163	1021034
Volume ('000 sme)	71256	103156	296459
Price US$/piece	14.72	10.37	9.07

Source: Complied from Office of Textile & Apparel, U.S. Dept. of Commerce

viii) Suppliers of women's and girls' wool coats (category 435)

Italy is the leader in supplying this category in terms of value due to its target to high end customer (average price US$ 103.79) which is much higher than average import price US$ 35.68. In terms of volume Dominican Republic followed by India, Guatemala and Ukarine are other major suppliers to US market in this category. The average earnings for Dominican Republic (US$ 25.45) and Hong Kong (US$ 34.97) are much lower than average price of US imports from Italy. This category is 10th in value and 12 th in volume imports from US. India ranks 3rd in value and 2nd in volume with average price of $19.23 (Table 3.22) indicating India's lower end target for this category.

Table 3.22: Imports of Women's and Girls' wool coats

	1990	2000	2004
India			
Value (US$ '000)	9	7707	21541
Volume ('000 sme)	2	2112	4218
Price (US$/piece)	16.91	13.71	19.23
All US imports			
Value (US$ '000)	238447	496295	599630
Volume ('000 sme)	21074	60538	63153
Price US$/piece	42.52	30.81	35.68

Source: Complied from Office of Textile & Apparel, U.S. Dept. of Commerce

The Categories in which India's position is not significant in US market along with leading suppliers are:

ix) Women's and girls cotton trousers (category 348)

Table 3.23 show Mexico ranks first followed by Hong Kong, China, Philippines and Turkey in this category) .The average import price is US$ 5.88 This category has seen a good growth rate in last decade. The average earnings for imports from Mexico is $7.01 while for China US$ 9.89 and from Turkey it is US$ 4.88.This category is I rank in import from US in volume and II in value indicating its importance in US imports. India despite of its strong cotton fiber base does not figure in top-10 counties from import takes place.

Table 3.23: Imports of Women's and Girls cotton trousers

	1990	2000	2004
Mexico			
Value (US$ '000)	83075	1548359	1310259
Volume ('000 sme)	18913	301457	224904
Price (US$/piece)	5.45	6.38	7.01
Hong Kong			
Value (US$ '000)	395637	401227	552287
Volume ('000 sme)	58758	60989	77508
Price (US$/piece)	8.36	8.17	8.99
China			
Value (US$ '000)	n/a	102042	161555
Volume ('000 sme)	n/a	11617	18406
Price (US$/piece)	n/a	10.91	9.89
Philippines			
Value (US$ '000)	46182	155358	194175
Volume ('000 sme)	8844	25755	33559
Price (US$/piece)	6.48	7.49	6.70
Turkey			
Value (US$ '000)	60708	190406	202603
Volume ('000 sme)	17359	51627	46058
Price (US$/piece)	4.34	4.58	4.88
All US imports			
Value (US$ '000)	1490007	4860779	6332.227
Volume ('000 sme)	296681	960702	1275377
Price (US$/piece)	6.24	6.28	5.88

Source: Complied from Office of Textile & Apparel, U.S. Dept. of Commerce

x) Suppliers of men's and boys' cotton trousers (category 347)

Mexico, Dominican Republic, Vitenam, Hong Kong followed by Guatemala and Cambodia are key suppliers of this category (Table 3.24). The average import price is US$ 6.44 and the trade has increased by around 8.4 percent in last decade. Hong Kong ($ 10.78) is catering to higher end of the market in this category. Category 347 is one of large volume categories being imported (2nd in import from US in volume and 6th in value). Here too, India is not present in top-10 countries from where import takes place.

Table 3.24: Imports of Men's and Boys' cotton trousers

	1990	*2000*	*2004*
Mexico			
Value (US$ '000)	110856	1669595	1469949
Volume ('000 sme)	24490	278541	243719
Price (US$/piece)	5.62	7.44	7.42
Dominican Republic			
Value (US$ '000)	161135	507724	449711
Volume ('000 sme)	34590	93466	81729
Price (US$/piece)	5.78	6.74	7.60
Hong Kong			
Value (US$ '000)	219747	279619	247528
Volume ('000 sme)	31328	33554	29198
Price (US$/piece)	8.71	10.35	10.78
Guatemala			
Value (US$ '000)	22719	114384	164547
Volume ('000 sme)	6849	20146	27715
Price (US$/piece)	4.12	7.05	7.47
Vietnam			
Value (US$ '000)	n/a	1422	147375
Volume ('000 sme)	n/a	688	30555
Price (US$/piece)	n/a	2.57	4.98
All US imports			
Value (US$ '000)	1339387	5014495	5023370
Volume ('000 sme)	267374	907411	949400
Price (US$/piece)	6.22	6.86	6.44

Source: Complied from Office of Textile & Apparel, U.S. Dept. of Commerce

The average import price is US$ 6.44 and the trade has increased by around 8.4 percent in last decade. Hong Kong ($ 10.78) is catering to higher end of the market in this category. Category 347 is one of large volume categories being imported (2nd in import from US in volume and 6th in value). Here too, India is not present in top-10 countries from where import takes place.

xi) Suppliers of women's and girls' cotton knit shirts (category 339)

Mexico, Guatemala followed by Honduras, Hong Kong is key suppliers for this category (Table 3.25). The average price of import is US$ 3.04. Hong Kong is catering to higher end of market with average earnings of US$ 6.37, which is much more than earnings of Mexico (US$ 1.99) & Guatamela (US$ 2.20) in the same category. The average price of imports from Honduras ($1.92) $ El Salvador ($1.90) are also observed for this category import from US. Category 339 is 3rd in value and 4th in volume in imports from US. Despite of strong cotton base India does not figure in key players in US market for this category.

Table 3.25: Imports of Women's and Girls' cotton knit shirts

	1990	*2000*	*2004*
Mexico			
Value (US$ '000)	12579	581336	540591
Volume ('000 sme)	2036	52654	135353
Price (US$/piece)	3.09	3.36	1.99
Guatemala			
Value (US$ '000)	14445	353515	729630
Volume ('000 sme)	3313	52654	135790
Price (US$/piece)	2.18	3.36	2.20
Honduras			
Value (US$ '000)	3469	329300	408749
Volume ('000 sme)	655	71089	105958
Price (US$/piece)	2.65	2.32	1.92
Hong Kong			
Value (US$ '000)	226409	311806	300604
Volume ('000 sme)	22179	24020	23574
Price (US$/piece)	5.10	6.49	6.37
Turkey			
Value (US$ '000)	57172	206149	192923
Volume ('000 sme)	5487	28664	27412
Price (US$/piece)	5.21	3.60	3.51
All US imports			
Value (US$ '000)	1130676	4359686	6095882
Volume ('000 sme)	134187	644061	1001336
Price (US$/piece)	4.24	3.38	3.04

Source: Complied from Office of Textile & Apparel, U.S. Dept. of Commerce

xii) Suppliers of men's and boys cotton coats (category 334)

In value terms, China followed by Mexico and Pakistan, Hong Kong, Sri Lanka are leading suppliers of this category to US market (Table 3.26) while in terms of volume Pakistan is the leading supplier followed by China, Bangladesh, Thailand, Mexico. The average import price US$ 9.77 is much higher than average import price from Pakistan i.e. US$5.89. Mexico ($18.71) and Hong Kong($15.60) are other countries with higher per piece earnings. The average price of import from China is US$ 15.42. This category is 11th in value and 10th in volume imports from US.

Table 3.26: Imports of Men's and Boy's cotton coats

	1990	2000	2004
China			
Value (US$ '000)	47203	67129	66069
Volume ('000 sme)	8898	11296	12547
Price (US$/piece)	15.25	17.09	15.13
Hong Kong			
Value (US$ '000)	51489	51533	34408
Volume ('000 sme)	8682	9643	6488
Price (US$/piece)	17.05	15.36	15.24
Mexico			
Value (US$ '000)	1577	29835	65410
Volume ('000 sme)	477	6687	10511
Price (US$/piece)	9.50	12.83	17.89
Pakistan			
Value (US$ '000)	3037	24978	54919
Volume ('000 sme)	1038	11991	23595
Price (US$/piece)	8.41	5.99	6.69
Sri Lanka			
Value (US$~ '000)	15785	28917	46762
Volume ('000 sme)	4531	6176	12388
Price (US$/piece)	10.02	13.46	10.85
All US imports			
Value (US$' 000)	284046	386789	579182
Volume ('000 sme)	54235	99307	164888
Price US$/piece	15.06	11.20	10.09

Source: Complied from Office of Textile & Apparel, U.S. Dept. of Commerce

3.3.2 European Union market

China (13.9 percent share), Turkey (12.9 percent) followed by India (9.3 percent), Pakistan (7 percent) & Czech Republic (5.7 percent) are top five suppliers of MFA Textiles to EU market(Table 3.27). India ranks IIIrd in value as well as volume in imports from EU.Switzerland,

US, S.Korea, Poland and Indonesia are other leading suppliers of MFA textiles to EU. The average prices of imports are Euro 4.56 /pc while India has 3.84 Euro/pc. The average prices have decreased in last decade for India.

Table 3.27: Imports of MFA textiles

	1990	2000	2003
China			
Value (Euro '000)	546	1767	2201
Volume ('000 pieces)	120	336	523
Price (Euro/piece)	4.55	5.26	4.21
Turkey			
Value (Euro '000)	723	1841	2038
Volume ('000 pieces)	150	356	417
Price (Euro/piece)	4.82	5.17	4.89
India			
Value (Euro '000)	634	1690	1472
Volume ('000 pieces)	139	383	383
Price (Euro/piece)	4.56	4.41	3.84
Pakistan			
Value (Euro '000)	501	981	1111
Volume ('000 pieces)	129	232	307
Price (Euro/piece)	3.88	4.23	3.62
Czech Republic			
Value (Euro '000)	140	727	898
Volume ('000 pieces)	35	157	191
Price (Euro/piece)	4.00	63	4.70

Source: Complied from EUROSTAT, CIRFS, CITH, EURATEX

China followed by Turkey, Romania, Bangladesh, Tunisia are top five suppliers of MFA Clothing to EU market (Table 3.28). Macau, India, Hongkong, Poland, Indonesia are other leading suppliers of MFA clothing to EU. India has 7th rank in EU imports with 4.6 percent (value) & 5.2 percent (volume) of total imports from EU. China has 17.7 percent (value) & 23.6 percent (volume) share in imports from EU followed by Turkey (14 percent value and 12.3 percent in volume) imports of EU. The average price of imports has become 14.39 Euro/pc from 17.08 Euro/pc during 1990-2003. The prices have been Euro 17.26 to 12.90 /pc for India in corresponding period. The prices from china (Euro 10.82), Bangladesh (Euro 8.02) & Indonesia(Euro 12.68) are much lower than aveage import price to EU. Poland and Romania have more than Euro 20/pc for supplies to EU market.

Table 3.28: Imports of MFA Apparel

	1990	2002	2003
China			
Value (Euro '000)	1451	7910	8572
Volume ('000 pieces)	129	602	792
Price (Euro/piece)	11.25	13.14	10.82
Turkey			
Value (Euro '000)	1846	6696	7098
Volume ('000 pieces)	107	377	413
Price (Euro/piece)	17.25	17.76	17.19
Romania			
Value (Euro '000)	329	3497	3498
Volume ('000 pieces)	19	168	173
Price (Euro/piece)	17.32	20.82	20.22
Bangladesh			
Value (Euro '000)	235	2682	3046
Volume ('000 pieces)	26	276	380
Price (Euro/piece)	9.04	9.72	8.02
India			
Value (Euro '000)	863	2204	2244
Volume ('000 pieces)	50	159	174
Price (Euro/piece)	17.26	13.86	12.90
All extra – EU			
Value (Euro '000)	17835	48617	48323
Volume ('000 pieces)	1044	2992	3359
Price (Euro /piece)	17.08	16.25	14.39

Source: Complied from EUROSTAT, CIRFS, CITH, EUROTEX

The analysis of EU imports reflects (Table 3.29) that India is having VIII rank. China, Turkey, Romania, Tunisia, Bangladesh, Morocco and Hong Kong are having better performance than India in EU market.

India is having II rank in Ladies dresses, Gent's shirts and Women's suit category. Besides it, India has presence in other categories namely ladies blouse (III rank) and T-shirts (IV rank) and Ladies skirts (VI rank) while in all other categories, India doesn't figure in the list of top 10 countries from where EU import is taking place. The average price of imports from India for most of the items is lower than overall average import price. Besides it, India doesn't figure as supplier in categories with high average price.

Table 3.29: Suppliers of Apparel to EU

Category/Country	Rankings				
	I	II	III	IV	V
	China	Turkey	Romania	Tunisia	Bangladesh
Ladies Blouses	Turkey	Romania	India	HK	Poland
Gents Shirts	Bangladesh	India	Romania	Turkey	HK
Ladies Dresses	China	India	Turkey	Romania	Morocco
Ladies Skirts	China	Turkey	Romania	Morocco	Tunisia
Trousers	Turkey	Tunisia	Romania	Morocco	HK
T-Shirts	Turkey	Bangladesh	China	India	Morocco
Pullovers	Turkey	Bangladesh	HK	Romania	Indonesia
Women's Suit	China	India	Turkey	Romania	Poland
Women's Overcoat	China	Romania	Poland	Turkey	Morocco

Source: Complied from EUROSTAT, CIRFS, CITH, EURATEX

The detailed analysis of India's competitive position for select categories in EU market is as follows:

i) T-shirts (category 4)

India ranks third in volume and fourth in value terms (Table 3.30) in this category in EU market. The average price of imports from India is Euro 2.86 and is marginally higher than average price (Euro 2.65). Turkey is the leading supplier with 30% (value) & 21.4% (vol) in EU imports with Euro 3.71 average price of imports. Bangladesh and China have higher share in import from EU in this category. Mauritius, Morocco, Hong Kong are other leading destinations of EU imports China is price leader with Euro 4.12 while Bangladesh has average price of only Euro 1.34 indicating their respective earnings so target market too.

Table 3.30: Imports of T-shirts

	1990	2002	2003
Turkey			
Value (Euro '000)	244180	1428547	1683109
Volume ('000 pieces)	56773	454042	454042
Price (Euro/piece)	4.30	3.71	3.71
Bangladesh			
Value (Euro '000)	57495	776600	776600
Volume ('000 pieces)	57335	581589	581589
Price (Euro/piece)	1.00	1.34	1.34
China			
Value (Euro '000)	54721	450820	356066
Volume ('000 pieces)	46300	96083	86358
Price (Euro/piece)	1.18	4.69	4.12
India			
Value (Euro '000)	72262	318753	312098
Volume ('000 pieces)	31469	115941	109.292
Price (Euro/piece)	2.30	2.75	2.86
All Extra – EU			
Value (Euro '000)	1358867	5352526	5619399
Volume ('000 pieces)	580222	1841440	2124180
Price (Euro /piece)	2.34	2.91	2.65

Source: Complied from EUROSTAT, CIRFS, CITH, EURATEX

ii) Women's blouses (category 7)

India ranks I in volume terms and II in value terms in this category preceded by Turkey. The trade has increased by around 4 percent in last decade. The average prices of import from India is Euro 3.90 which is much lower than average import price in Europe i.e. Euro 5.01(Table 3.31) indicating lesser than average price realization by exports from India. Romania, Hong Kong, Poland, Bulgaria , Morocco, China and Bangladesh are other suppliers for this category. Bulgaria & Bangladesh (Euro 2.43 /pc) target to lower price end of market while China (Euro 7.10) is targeting to upper price bracket in EU market.

Table 3.31: Imports of Women's Blouses

	1990	2000	2003
Romania			
Value (Euro '000)	17555	181058	300042
Volume ('000 pieces)	1597	29319	56706
Price (Euro/piece)	10.99	6.18	5.29
Turkey			
Value (Euro '000)	109867	219318	334357
Volume ('000 pieces)	14559	39342	57088
Price (Euro/piece)	7.55	5.57	5.86
Hong Kong			
Value (Euro '000)	225362	256435	180281
Volume ('000 pieces)	30665	32701	28047
Price (Euro/piece)	7.35	7.84	6.43
India			
Value (Euro '000)	190617	200603	331223
Volume ('000 pieces)	49811	52981	84939
Price (Euro/piece)	3.83	3.79	3.90
Poland			
Value (Euro '000)	59617	187982	152001
Volume ('000 pieces)	6102	30633	28541
Price (Euro/piece)	9.77	6.14	5.33
All Extra – EU			
Value (Euro '000)	1143147	2202119	2281709
Volume ('000 pieces)	204493	417491	455814
Price (Euro / piece)	5.59	5.27	5.01

Source: Complied from EUROSTAT, CIRFS, CITH, EURATEX

iii) Men's shirts (category 8)

In men's shirt (Category 8) India ranks second in import from EU preceded by Bangladesh. Romania, Turkey, Hongkong and China are other key players in this category. Table 3.32 shows that trade has increased by 3.8 percent in last decade in value terms but average price (Euro 5.55) is slightly higher than the price in 1990 (Euro 5.12). The average import price is Euro 5.55. Bangladesh (Euro 2.7) is leading in imports while Tunisia (Euro 10.17), Romania (Euro 9.16), Turkey (Euro 8.30), China (Euro 5.95) are other competitors of India in this category.

Table 3.32: Imports of Men's shirts

	1990	2000	2003
Bangladesh			
Value (Euro '000)	119491	409240	330791
Volume ('000 pieces)	48422	136774	122302
Price (Euro/piece)	2..47	2.99	2.70
India			
Value (Euro '000)	167076	203023	250770
Volume ('000 pieces)	32271	39195	55371
Price (Euro/piece)	5.18	5.18	4.53
China			
Value (Euro '000)	34322	123540	103435
Volume ('000 pieces)	9744	18344	17380
Price (Euro/piece)	3.52	6.73	5.95
All Extra – EU			
Value (Euro '000)	1467970	2352690	2285448
Volume ('000 pieces)	286866	378678	411734
Price (Euro/piece)	5.12	6.21	5.55

Source: Complied from EUROSTAT, CIRFS, CITH, EURATEX

iv) Women's dresses (category 26)

India ranks I in in this category in EU imports. The average import price from India is Euro 4.70 while average import price for overall imports is Euro 7.04 indicating that India's exports are targeted to lower end of the market in this category as well (Table 3.33). China has II rank in value and VI in volume in this category with average price of Euro 18.23 indicating upper end target. Turkey, Romania, Morocco and Hong kong, Tuniasia, Poland are other key players in this market. Sri Lanka (Euro 4.29/pc) is having 9th rank in imports and too target lower price end.

Table 3.33: Imports of Women's dresses

	1990	2000	2003
China			
Value (Euro'000)	10036	114058	103543
Volume ('000 pieces)	2026	5090	5681
Price (Euro/piece)	4.95	22.41	18.23
India			
Value (Euro'000)	46152	123832	106606
Volume ('000 pieces)	7291	1887	22536
Price (Euro/piece)	6.33	6.56	4.70
Turkey			
Value (Euro'000)	66029	104053	75215
Volume ('000 pieces)	13280	17103	10959
Price (Euro/piece)	4.97	6.08	6.86
Sri Lanka			
Value (Euro'000)	7408	40023	19804
Volume ('000 pieces)	1286	12571	4661
Price (Euro/piece)	5.76	3.18	4.25
All Extra – EU			
Value (Euro'000)	447459	1038469	748783
Volume ('000 pieces)	72835	137601	106404
Price (Euro/piece)	6.14	7.55	7.04

Source: Complied from EUROSTAT, CIRFS, CITH, EURATEX

v) Women's skirts (category 27)

In Women's skirts India ranks fifth in EU imports (Table 3.34). The average import price (Euro 4.74) is much lower than average import price for this category i.e. Euro 5.95. The average price for imports from India has reduced from 5.97 Euro per piece to 4.74 Euro per piece during 1990-03 while average import price has only reduced from 7.72 to 5.95 Euro per piece. China, Turkey, Romania, Morocco & Tunisia are other leading suppliers of this category. The average price of imports from China(Euro 4.08) is on lower side while the other leading suppliers Romania, Turkey (Euro 7.54) are having relatively much higher earnings per piece in EU import of this category.

Table 3.34: Imports of Women's skirts

	1990	*2000*	*2003*
China			
Value (Euro'000)	19417	241186	313027
Volume ('000 pieces)	4154	52350	76638
Price (Euro/piece)	4.67	4.61	4.08
Romania			
Value (Euro'000)	19962	125723	189114
Volume ('000 pieces)	1890	38415	25082
Price (Euro/piece)	10.56	3.27	7.54
Turkey			
Value (Euro'000)	53921	125723	189114
Volume ('000 pieces)	7448	38415	25082
Price (Euro/piece)	7.24	3.27	7.54
India			
Value (Euro'000)	50032	69693	85940
Volume ('000 pieces)	8386	13961	18147
Price (Euro/piece)	5.97	4.99	4.74
All Extra – EU			
Value (Euro '000)	576803	1202232	1342434
Volume ('000 pieces)	74696	280480	225609
Price (Euro/piece)	7.72	4.29	5.95

Source: Complied from EUROSTAT, CIRFS, CITH, EURATEX

The Categories in which India's position is not significant in imports from EU along with leading suppliers are:

vi) Trousers (category 6)

This category is I in value and II in volume import from EU, Turkey, Tunisia followed by Romania and Morocco, Hong Kong are leading suppliers of trousers (category 6) to EU market (Table 3.35). India does not figure in top-10 countries in this category. The average price for this category is Euro 7.90 while the countries i.e. Turkey (Euro 9.66), Tunisia (Euro 10.85), Morocco (Euro 8.38),Romania(Euro 9.73)per piece from this category. Bangladesh (Euro 3.91) & Pakistan (Euro 5.0), China (Euro 8.23) per piece are other key countries in this category for EU imports.

Table 3.35: Imports of trousers

	1990	2000	2003
Turkey			
Value (Euro '000)	186115	903254	1370464
Volume ('000 pieces)	19394	96466	141853
Price (Euro/piece)	9.60	9.36	9.66
Tunisia			
Value (Euro '000)	313417	916266	945035
Volume ('000 pieces)	41390	101919	87111
Price (Euro/piece)	7.57	8.99	10.85
Morocco			
Value (Euro '000)	247377	726413	770004
Volume ('000 pieces)	34754	97686	91895
Price (Euro/piece)	7.12	7.44	8.38
Romania			
Value (Euro '000)	39064	607198	890979
Volume ('000 pieces)	3910	60841	91546
Price (Euro/piece)	9.99	9.98	9.73
Bangladesh			
Value (Euro '000)	19563	460928	729210
Volume ('000 pieces)	5717	94488	186467
Price (Euro/piece)	3.42	4.88	3.91
All Extra – EU			
Value (Euro '000)	2366863	7334270	8582043
Volume ('000 pieces)	309470	862063	1086118
Price (Euro/piece)	7.65	8.51	7.90

Source: Complied from EUROSTAT, CIRFS, CITH, EUROTEX

vii) Women's overcoats (category 15)

This category is V in value and VII in volume imports from EU. China, Romania, Poland followed by Turkey and Morocco are leading suppliers (Table 3.36) of this category. The average import price of this category is Euro 16.78. The average price of import for China (top in this category) is Euro 12.64. While from Poland (3rd in imports) it is 22.63 Euro per piece & Romania (2nd in EU imports) is 20.08 Euro/pc. The average import price from Hungary (9th in imports) is Euro 43.47/pc.

Table 3.36: Imports of Women's overcoats

	1990	2000	2003
China			
Value (Euro' 000)	95742	303470	361619
Volume ('000 pieces)	7364	76457	28619
Price (Euro/piece)	13.00	3.97	12.64
Romania			
Value (Euro'000)	41130	211362	273498
Volume ('000 pieces)	2006	10915	13622
Price (Euro/piece)	20.50	19.36	20.08
Poland			
Value (Euro'000)	65677	170564	130702
Volume ('000 pieces)	2528	6675	5775
Price (Euro/piece)	25.98	25.55	22.63
Turkey			
Value (Euro'000)	61961	131782	126281
Volume ('000 pieces)	3633	10009	10038
Price (Euro/piece)	17.06	13.17	12.58
Morocco			
Value (Euro'000)	35113	88392	102181
Volume ('000 pieces)	2922	7705	7838
Price (Euro/piece)	12.02	11.47	13.04
All Extra – EU			
Value (Euro'000)	830108	1768882	1767834
Volume ('000 pieces)	41769	161188	105356
Price (Euro/piece)	19.7	10.97	16.78

Source: Complied from EUROSTAT, CIRFS, CITH, EURATEX

3.4 SUMMARY

It can be summarized that India is catering to rather lower end of the market in key destination markets of apparel i.e. US and EU. In imports from US, India has fifth rank while in EU it has eighth rank in total imports. There are many categories where India does not have any significant presence or there are categories where the earnings for exports from India are rather low. The reason for above may be lying in relative competitiveness of India in world market. The analysis for imports of US market and India's competitive position in US imports clearly indicate that India doesn't have a significant share in the categories namely Women's and Girl's cotton trousers (Category 348), Women's and Girl's cotton knit shirts (Category 339), Men's and Boys' cotton trousers (Category 347), Women's and Girls' wool coats (Category 435) and Men's and Boys' cotton coats (Category 334). This is important to note that category 348 with average price $ 5.95 is having highest imports in terms of volume while category 339 is third in value terms imports from US showing significant importance of these categories in US market. This coupled with the fact that India doesn't have a significant presence in US market in these categories despite of its rich cotton based. It can infer that category 348, category 339, category 347 and category 334 are potential new categories for Indian garment exporters for US market.

The analysis of imports of EU and India's competitive position in EU apparel trades that India does not have a presence in categories namely Trousers (category 6) and Women's overcoats (Category 15). Out of these categories 6 is highest in terms of import value from EU while category 15 is fifth in rank in value terms imports. The average price for category 6 is Euro 8.33 while for category 15 it is Euro 17.90. It can be inferred that trouser (category 6) and women's over-coats (category 15) are potential new categories for Indian garment exporters for EU market. Indian apparel exporters cater to lower end of market in most of the categories in key destination market i.e. US, EU. The need is to target categories with higher realization and de-mand in destination market.

Chapter 4

Competitiveness of Indian Apparel Industry: Study of Perceptions

According to ATC, beginning 1st January 1995, all textile and apparel products that had been hitherto subjected to MFA-quota is to be integrated over a period of ten year. The dismantling of the quota regime represents both an opportunity as well as a threat. An opportunity because markets will no longer be restricted, a threat because markets will no longer be guaranteed by quotas, and even the domestic market will be open to competition. From 1st January 2005, therefore all textile and apparel products are to be traded internationally without quota restrictions. This impending reality brings the issue of competitiveness to the fore for all firms in the textile and apparel sector, including those in India. It is imperative to understand the true competitiveness of Indian textile and apparel firms in order to make an assessment. The competitiveness of the Indian textiles and apparel industry has assumed great significance after 31st December 2004, when quota free trade has become reality.

The chapters 4th-6th of this book are based on findings of primary data collected through survey of apparel exporters, fabric manufacturers and buying houses with structured questionnaires. Apparel exporters from all the major manufacturing centres have been covered in the survey of apparel exporters. Convenience & Judgmental sampling was the method used in all the three cases. The questionnaires were administered with (i) apparel exporters situated in major production centers in India like NCR region, Bangalore, Ludhiana, Tirupur and Chennai. (ii) fabric manufacturers from both mill and power loom Sector. The mill sector of Western and Southern India had been covered along with powerloom sector of Bhiwandi, Surat and Salem. (iii) buying houses situated at major trading centers like NCR region, Mumbai and Bangalore.The respondents among apparel exporters were drawn from among the top 500 apparel exporters registered with the Apparel Export Promotion Council (AEPC). About 30 percent of the sample belonged to the top 200 apparel exporters. The respondents among fabric manufacturers were drawn from the list of fabric manufacturers. Similarly the buying houses surveyed were selected from top–100 buying houses from the directory of buying houses.

In order to study the competitiveness of Indian textile (fabric) and apparel sector the variables in the apparel exporters questionnaire were grouped together under preference for manufacturing / marketing woven or knitted; preference for various sources of fabric and price points of various categories exported. The variables in the fabric manufactures' and buying house' questionnaire were also summarized in the similar manner for comparison. The perceptions are analysed to achieve the desired objectives.

The literature on perception indicates the importance it has in decision-making process. 'Perception' is defined as the process by which an individual selects, organizes and interprets stimuli into a meaningful and coherent picture of the world. The study of perception is largely the study of what we subconsciously add to or subtract from raw sensory inputs to produce our own private picture of the world. (leon G. schiffman & leslie lazar Kanuk, 2000). When we do make a decision on a purchase, we are responding not only to influences of various parameters but to our interpretations of them (Michael R. Solomon, 1999). Perception is a three step process that involves selecting, organizing and interpreting specific stimuli in a situation according to prior learning, activities interest, experience etc. It is a process and a pattern of response to stimuli or information inputs (Wolfgang J. Koschnick, 1995).

In order to ascertain the difference in importance given to various aspects while manufacturing/marketing or sourcing woven/knitted product and manufacturing greige/finished fabric, sourcing fabrics from various sources, the opinion regarding lesser focus on value added products; the weights derived from importance scale and their relative rankings were analyzed. The differences between response of the apparel exporters and the buying houses on common variables were analyzed through 'Weighted Performance Measurement'.

For analysis of all three-questionnaire(s); frequencies, weighted average performance and importance rankings are used to meet the specific objectives of the research. All the frequencies are converted into percentage of respondents and expressed upto two places of decimals and diagrammatically represented by line graph and radar diagram etc. The respondents were asked to indicate the importance or preference of parameters on a 5-point scale ranging from 1 (extremely low / least preferred) to 5 (extremely high/most preferred). The weighted average scores are computed and expressed upto two places of decimals; the perceptual distances are then graphically represented through radar diagrams. The perceptual distances show differences of 'group mean scores' of variables and are relative expression of preference towards parameters, which has no absolute value. Wherever, the responses are drawn from both (apparels exporters and buying houses) for same purpose; a radar diagram has been made to show the differences in their opinion. The perception maps are also drawn for showing the difference of perception of apparel exporters and buying houses towards various parameters considered while making a sourcing decision of fabrics. The respondents were also asked to indicate their degree of agreement for statements on a 5-point scale (5 for agree completely to 1 for disagree completely) about lesser sourcing / focus on high value added products. Besides it, they were also asked to indicate the importance of measures on a 5-point scale (1 for not at all important to 5 for extremely important) for initiatives to be taken for facing international competition in post MFA period. The weighted averages are computed and expressed upto two places of decimals shown with a radar diagram. Radar diagram or chart is a tool that provides a visual display of the current state or level of performance against various parameters. It represents areas of improvement graphically. The fabric manufactures were asked to indicate importance of parameters for manufacturing woven or knitted fabrics on a 5-point scale and weighted averages are computed and expressed upto two places of decimal and shown with a help of a radar diagram. The fabric manufacturers were also asked the reasons for lesser focus on high value items, importance of parameters to increase fabric exports and the importance of parameters for increasing import of fabric on 5-point scale and the weighted averages or frequencies were shown with help of diagrammatic representations.

To validate the findings from perception study regarding preferred source of fabric by apparel exporters and buying house and the reasons of lesser focus on manufacturing or sourcing of high value items; Factor analysis is used. The factor analysis is used for questions with large set of variables only to investigate the interrelations with all relevant variables. 'Factor analysis' is a statistical technique for classifying a large number of interrelated variables into a limited number of dimensions or factors. It is a useful method for constructing multiple-item scales, where each scale represents a dimension of a highly abstract construct. It is an efficient method for reorganizing the items a researcher is investigating into conceptually more precise groups of variable. Factor analysis seeks to identify a set of dimensions that is not readily observed in a large set of variables.

The research design and methodology comprehensively met the research objectives by having samples from apparel exporters, fabric manufacturers and buying houses in different region and different segments through in depths interviews with the help of structured questionnaire. The collection of information was done systematically to provide authentic findings from the study. The analysis of market share and growth rates of India in target markets was the empirical basis for appreciating the qualitative findings. With the use of frequencies, rankings and weighted or unweighted performance scores as the case may be, the perceptions are analyzed supported by tables and radar diagrams and the findings are presented in easily comprehensible manner.

In this chapter, the findings from the survey of apparels exporters and buying houses are discussed. The sample of research consisted of 200 apparel exporters with 80 percent apparel manufacturer - exporters and 20 percent merchant exporters. The respondents are drawn from top 500 exporters of India so as to fairly represent Indian apparel export industry.

The sample of buying houses consisted of 60 buying houses representing major buyers in destination (import) market. The analysis is carried out under different sub-themes and is in the following sequence:

1. Product (apparel) categories exported / sourced from India
2. Lesser focus on high value added products
3. Productivity of apparel firms
4. Reasons of exporting/ sourcing woven vs. Knitted apparel
5. Availability of fabric for apparel from various sources

The analysis of product categories being exported by apparel exporters or product categories sourced by buying houses representing importers and the average price (FOB) realized provide an in-depth understanding of target market in terms of categories and prices (FOB) for Indian apparel exporters. This alongwith reasons for lesser focus on export or sourcing of high value added products from India shall help in understanding the positioning of Indian apparel items and devising strategy to fill that gap; if Indian exporters intend to cater to upper end of market. The response from exporters shall also help in understanding productivity levels in Industry so as to have a view on competitiveness of industry in this respect. The availability of raw material i.e. fabric is a critical factor in determining competitiveness of industry. The perception of apparel exporters and buying houses for various sources of fabric availability for apparel industry studied in present research shall help in determining competitive position of Indian fabric vis-à-vis imported fabric.

4.1 AVERAGE PRICE FOR DIFFERENT PRODUCT CATEGORIES EXPORTED

India is exporting various apparel categories to different destination market and it is pertinent to examine the key categories being exported or sourced and price point at which the exporters are working or the price point at which the sourcing is being preformed by buying houses. Hence, it is necessary to study the response of apparel exporters as well as buying houses to understand key categories of apparel being manufactured in India. The responses of both the sets of respondents shall collectively provide an overview.

4.1.1 Response of apparel exporters

Table 4.1 shows percentage of respondents exporting various categories of apparels from India. The key categories being exported from India include Gent's shirts (51.03 percent of respondents) followed by T-shirts (48.97 percent of respondents), Ladies skirts (46.90 percent of respondents), Trousers (45.52 percent of respondents). Besides it, ladies dress (42.76 percent of respondents), Ladies blouse (40 percent of respondents) and Jackets are other categories being exported from India.

A detailed analysis of findings (Table 4.1 & Exhibit 4.1) indicates that the exports of Ladies blouses category is between price range ($4-$8) followed by price below $4. Gent's shirts ($4-$8) followed by below $4, Ladies dress ($4-$8) followed by ($8-$12), Ladies skirts ($4-$8) followed by below $4, Trousers ($4-$8) followed by below $4, T-shirts below ($4 and $4-$8) and Jackets ($4-$8 and $8-$12) are other key categories alongwith average FOBs for exports from India. It reflects an average FOB of below $8 in most of the categories with some categories even having concentration on below $4 price point.

Table 4.1: Realisation of Apparel categories exported

Product category	Below $4	$4—$8	$8—$12	$12—$16	Above $16	Percentage of respondents
Ladies blouses	20.69	60.34	8.62	1.72	0.00	40.00
Gents shirts	9.46	63.51	8.11	6.76	0.00	51.03
Ladies dresses	9.68	48.39	32.26	4.84	0.00	42.76
Ladies skirts	26.47	55.88	11.76	0.00	1.47	46.90
Trousers	21.21	51.52	16.67	6.06	0.00	45.52
T-shirts	46.48	38.03	7.04	1.41	1.41	48.97
Jackets	2.13	25.53	23.40	17.02	21.28	32.41
Any other*	20.83	41.67	20.83	4.17	0.00	16.55

*Any other includes Babies/Children wear, Intimate apparel, Sweater and Underwears etc.

Exhibit 4.1: Realisation of Apparel categories exported

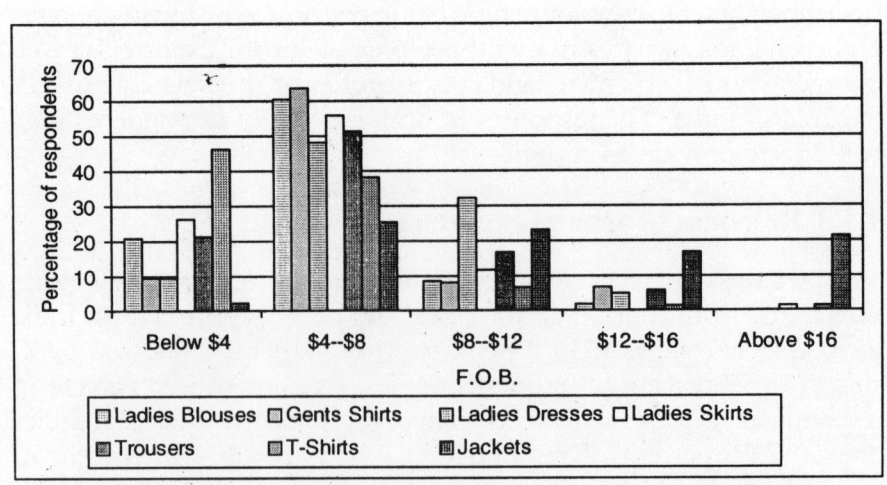

The price range of above $12 is found only for Jackets besides a small percentage for Trousers, Gents shirts and shorts and Ladies dresses. The categories (T-shirt, Ladies skirts, Trousers and Ladies blouses) have significant export at price below $4 indicating targeting to lower end of market for these catogories. It indicates that India primarily targets to lower end of the market in most of the categories being exported from India.

4.1.2 Response of buying houses

Response of the survey of buying houses shown in Table 4.2 & Exhibit 4.2 shows that Gents shirt (49 percent of respondents), Trousers (48 percent of respondents), ladies blouses (47 percent of respondents) and T-shirts (45 percent of respondents) are key categories being sourced from India followed by ladies skirts, ladies dresses and jackets.

The average price for ladies blouses being sourced by buying houses is below $ 8 with more concentration of responses for average price of ($ 4 - $ 8).

Gent's shirts is being sourced at average FOB ($4-$8) followed by below $4); Ladies dresses is having an average FOB ($8 – $12) followed by ($12 – $16)

Ladies Skirts is being sourced at average FOB ($4 - $8) followed by ($ 8 - $ 12). The average FOB of trousers being sourced from India is ($8-$12) followed by ($4-$8).

T-shirts ($4 - $8) and Jackets (above $12) are other categories being sourced. It indicates that the average FOB for most of the categories is ($4-$8) followed by below $4 and $8-$12 except for the Jackets where the average FOB is above $ 12.

Table 4.2: Realisation of Apparel categories sourced

Product category	Below $4	$4—$8	$8—$12	$12—$16	Above $16	Percentage of respondents
Ladies Blouses	31.91	44.68	17.02	2.13	4.26	47
Gents Shirts	22.45	53.06	16.33	4.08	4.08	49
Ladies Dresses	0.00	14.71	38.24	29.41	17.65	34
Ladies Skirts	5.41	51.35	32.43	5.41	5.41	37
Trousers	16.67	35.42	37.50	8.33	2.08	48
T-Shirts	20.00	53.33	20.00	4.44	2.22	45
Jackets	7.69	0.00	7.69	38.46	46.15	26

Exhibit 4.2: Realisation of Apparel categories sourced

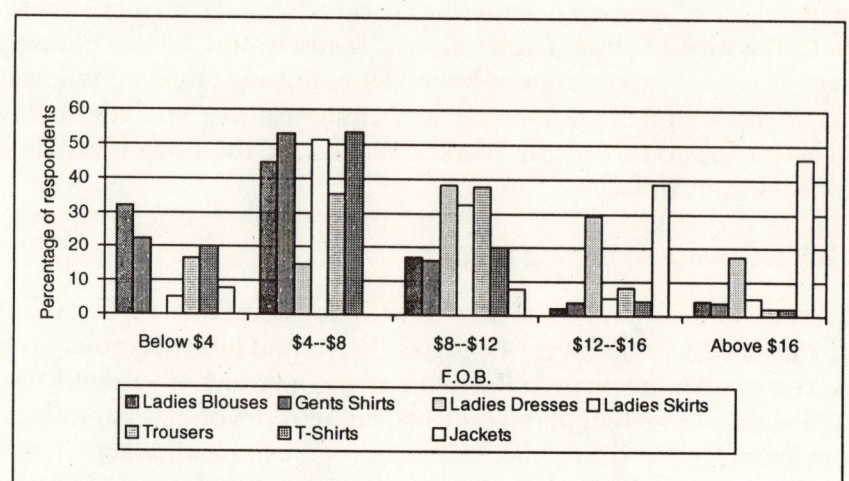

The response of apparel exporters and buying houses indicates that Gent's shirt, T-shirts, Trousers, Ladies skirts, Ladies blouses, Ladies dresses are key categories of apparels being exported or sourced from India. The export data indicate that these categories constitute around 70 percent of total exports from India.

The response of apparel exporters as well as buying houses indicates that in most of the apparel categories being exported the average FOB price is less than $8. There is sizeable number of respondents indicating even price realization below $4. The average FOB is above $8 for a few categories including trousers, ladies dresses and jackets where the quantity of exports or sourcing is low which is reflected by lesser percentage of respondents indicating that they manufacture or source these categories. The reasons may be lying in product categories being exported or target client in destination market and the material (fabric) being used in apparel leading to lower realization in terms of prices. In other words, for the most of the categories our exports are targeted to lower end of world apparel market.

4.2 LESSER FOCUS ON EXPORT OF HIGH VALUE ADDED PRODUCTS

As discussed earlier, India is having lesser focus on high value added products resulting in low FOB for most of the categories being exported. The response of apparel exporters and buying houses has been taken to know their perception of reasons of not targeting to high value segment in destination market. The perceptions are on a scale of 1-5 (1 being least important and 5 being most important). The weights are assigned and weighted average scores are shown in table and represented with the help of radar diagram (Table & Exhibit 4.3 and 4.4).

4.2.1 Response of Apparel Exporters

The response of apparel exporters indicates limited research and development facility, unfavorable cost competitiveness, non-availability of the quotas and India's image as basic product manufacturer and technological constraints as the key reasons behind not targeting to

Table 4.3: Lesser focus on high value added products

Statements	Average Score
India is primarily known for basic products	3.52
Req. design skills are not easily available	2.87
Raw material not readily available	2.92
Inadequacy of workers with required skills	2.77
Technological Constraints	3.47
Difficulty in importing raw material	3.37
Lesser demand in international market	2.54
Difficulty in locating the destination market	2.77
Unfavorable cost competitiveness	3.56
Quotas are not available	3.55
Limited finishing facilities	3.27
Limited R&D facility	3.82

Exhibit 4.3: Lesser focus on high value added products

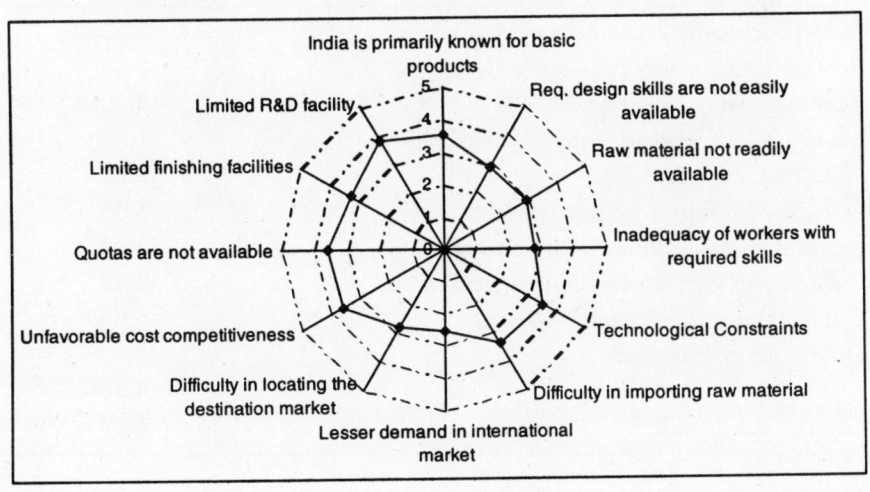

high value added segment. The other reasons include difficulty in importing raw material and limited finishing facility (Table 4.3 & Exhibit 4.3). The key reasons as indicated show lack of research and development facility and technological constraints leading to unfavorable cost competitiveness and giving India image as a producer of basic apparels. This coupled with availability of quotas play important role in apparel exporters not focusing on high value items in exports from India. The other reason i.e. lesser demand in international market, difficulty in locating destination market and inadequacy of workers with skills are not perceived to be important enough in determining focus of exporters on low value items.

4.2.2 Response of Buying Houses

The responses of survey of buying houses indicate that limited R&D facility, limited finishing facilities, unfavorable cost competitiveness and difficulty in importing raw material followed by India's image associated with basic products, availability of raw-material are the key reasons behind not sourcing much of value-added products from India. The other reasons include technological constraints in developing high value added products in India (Table 4.4 & Exhibit 4.4).

Inadequacy of workers with required skills, non-availability of design skills, difficulty in locating the destination market and lesser demand in international market are not the key reasons behind lesser sourcing of high value items from India.

Table 4.4: Lesser focus on high value added products

Statements	Average score
India is primarily known for basic products	3.63
Required design skills are not easily available	2.05
Raw material not readily available	3.50
Inadequacy of workers with required skills	1.89
Technological Constraints	3.21
Difficulty in importing raw material	3.66
Lesser demand in international market	2.03
Difficulty in locating the destination market	2.43
Unfavorable cost competitiveness	3.68
Quotas are not available	2.70
Limited finishing facilities	3.86
Limited R&D facility	3.97

Exhibit 4.4: Lesser focus on high value added products

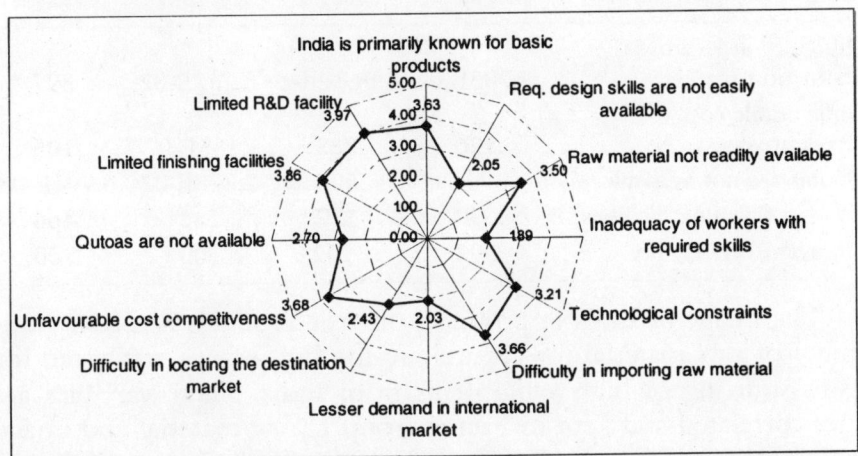

Statistical Analysis of Response of Buying Houses

The factor analysis of response of buying houses is undertaken to analyze the reasons of lesser sourcing of high value added products from India. Factor analysis using Principal Component Analysis (PCA) was carried out on the data collected, in order to bring out the salient features of the response of buying houses towards lesser focus on sourcing of high value items. The results of factor analysis with SPSS package are shown in Table 4.5.

Table: 4.5: Factor groups for lesser focus on high value added products

Parameters	Group 1	Group 2	Group 3	Group 4
India is primarily known for basic products	.781	-.123	-1.766E-02	4.826E-02
Req. design skills are not easily available	.244	-.763	.318	5.954E-02
Raw material not readily available	.760	4.528E-02	1.092E-02	-.183
Inadequacy of workers with required skills	9.758E-03	-.269	.667	-4.553E-02
Technological constraints	.449	4.707E-02	.491	-.343
Difficulty in importing raw material	.699	.254	-.384	-.302
Lesser demand in international market	-6.009E-02	6.302E-02	.830	9.038E-02

(Contd.)

Parameters	Group 1	Group 2	Group 3	Group 4
Difficulty in locating the destination market	-8.703E-04	4.068E-02	-3.757E-02	.897
Unfavorable cost competitiveness	.126	.883	5.185E-02	.105
Quotas are not available	-.716	.503	-2.378E-02	6.994E-03
Limited finishing facilities	.639	.217	.182	.466
Limited R&D facility	.592	-.102	.105	.250

The image of India as producer of basic products and non availability of raw material for apparel are the key reasons attributed for lesser sourcing of high value items from India. These variables are inter-correlated and kept in Factor group I (raw material and image functions). The factor loading for the variable *"India is primarily known for basic products"* is 0.781 while the variable *"raw material not readily available"* is having factor loading of 0.760 & *"the variable difficulty in importing raw material"* has factor loading of 0.699. It shows (Table 4.6) that these variables which are grouped under *"raw material and image function"* are key factors leading to lesser sourcing of high value added products from India.

The total variance attributable to the first factor group is 27.49 percent. The total variance for factor group II is 44.73 percent, 57.04 percent, for group III and for group IV is 67.39 percent.

Table 4.6: Reasons for lesser focus on high value added products

Variables	Factor loading	Factor group title (I)
India is primarily known for basic products	0.781	Raw material and
Raw material not readily available	0.760	image function
Difficulty in importing raw material	0.699	

Table 4.7 indicate that factor loadings for variables i.e. difficulty in locating the destination market (Factor loading 0.897), Unfavorable cost competitiveness (Factor Loading 0.883) and lesser demand in international market (Factor loading 0.830) are high in factor groups II, III and IV. This indicates market and cost related functions as important factors behind lesser sourcing of high value items from India. The Indian manufacturers find difficulty in locating destination market and are lesser competitive due to cost factor in world market.

**Table 4.7: Reasons for lesser focus on high value
added products**

Variables	Factor loading	Factor group title
Difficulty in locating the destination market	0.897	
Unfavorable cost competitiveness	0.883	Market and cost functions
Lesser demand in international market	0.830	

4.2.3 Perception Difference between Buying Houses and Apparel Exporters

The analysis of reasons of lesser focus on high value products from India has been discussed in earlier section. The difference in the perception of buying houses and apparel exporters for lesser focus on high-value products from India is depicted (Table 4.8 & Exhibit 4.8). There has been a similar response regarding reasons attributing lesser focus on high value items on the parameters namely limited R&D facility, India is primarily known for basic products, unfavorable cost competitiveness, difficulty in importing raw material and technological constraints associated with high value items. However, there is a difference of opinion in few parameters whereas the buying houses feel that the raw material, finishing facilities are not easily available for high value items in India. On the other side, the apparel exporters attribute the lesser focus on high value items to non-availability of quotas, lesser demand in international market. There is a consensus among exporters and buying houses that required design skills are available in India. Besides it, locating the destination market and lesser demand in international market are not the reasons of lesser focus on export or sourcing of high value items.

In nutshell, the reason for lesser focus on high value item exports from India include the image (perception) of India as producer of basic items besides limited research as development facility which is much required for targeting high end of market. The other factors responsible are unfavourable cost competitiveness due to availability of raw material required, limited finishing facilities and Technological constraints. A significant finding indicated with relatively low score for required design skills are not available and inadequacy of workers until required skills can be explained with fact that there is a consensus that India has required design skills for high value added products that requiste skills are available in India for targeting and catering the market

Table 4.8: Perception difference between buying houses and apparel exporters

Statements	Apparel exporters	Buying houses
India is primarily known for basic products	3.52	3.63
Required design skills are not easily available	2.87	2.05
Raw material not readily available	2.92	3.50
Inadequacy of workers with required skills	2.77	1.89
Technological Constraints	3.47	3.21
Difficulty in importing raw material	3.37	3.66
Lesser demand in international market	2.54	2.03
Difficulty in locating the destination market	2.77	2.43
Unfavourable cost competitiveness	3.56	3.68
Quotas are not available	3.55	2.70
Limited finishing facilities	3.27	3.86
Limited R&D facility	3.82	3.97·

All numbers shows average score.

Exhibit 4.8: Perception difference between buying houses and apparel exporters

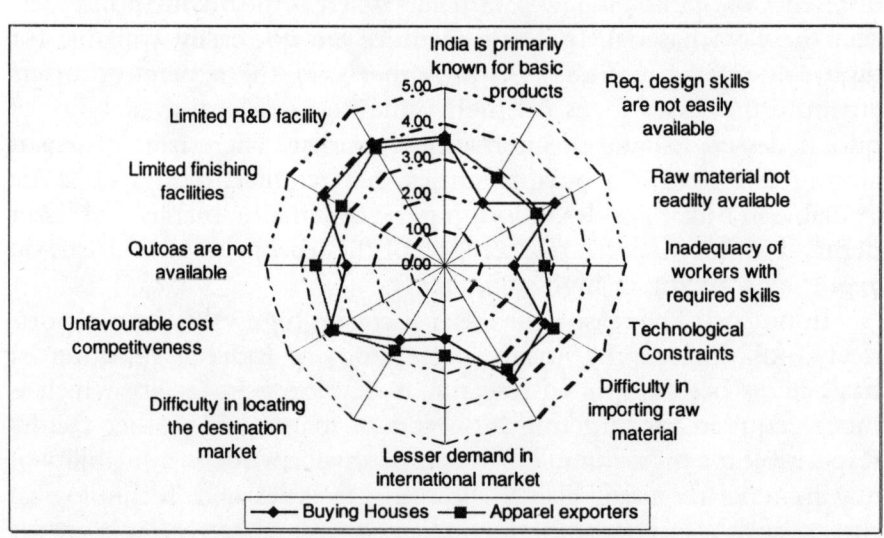

of high value added products. It is also noticeable that there is not much response to statement that there is lesser demand of high value items in international market. In other words, it can be interpreted that although there is a demand in international market for high value added products but due to limited research and development; finishing facilities and India's image as producer of basic items; there is lesser focus on high value added products in Indian apparel industry.

4.3 PRODUCTIVITY OF APPAREL FIRMS

In the post-quota period, the industry is set to witness a tough competition with neighboring countries with high productivity. As discussed earlier, performance of Indian apparel industry in terms of productivity is rather poor in comparison to other competing countries. The productivity is one of the key criteria for determining competitiveness in world apparel trade. The response of apparel exporters indicate the productivity (pieces/machine/day) for various categories being manufactured in India. For Ladies blouses, Gent's shirts, Ladies skirts, Trousers categories productivity is 8-12 pieces/machine/day; while for ladies dresses the productivity is between 4-12 pieces/machine/day. For T-shirts majority of respondents have indicated productivity as above16 pieces/machine/day. In Jackets category, the productivity is 4-8 followed by 8-12 pieces/machine/day (Table 4.9 & Exhibit 4.6).

In Table 4.9, the column percent of respondents indicate the percentage of respondents (exporters) exporting a particular category. e.g. 51.03 percent of respondents export gents shirts, 48.97 percent of respondents T-shirts followed by 46.9 percent of respondents. T-shirts followed by 46.9 percent respondent exporting Ladies skirts etc.

As already discussed and indicated in Table 4.15 the productivity of most of the items in India is less than 10.3 pieces/per machines/per day except for T-shirts and Jackets. The similar response is observed in survey of apparel exporters. It is a critical factor having adverse impact

Table 4.9: Productivity of apparel firms (pieces/machine/day)

Product Category	No of Pieces						Percentage of respondents
	< 4	4—8	8—12	12—16	>16	NA	
Ladies blouses	1.67	15.00	46.67	8.33	16.67	11.67	40.00
Gents shirts	0.00	12.99	40.26	20.78	18.18	7.79	51.03
Ladies dresses	4.62	33.85	33.85	3.08	18.46	6.15	42.76
Ladies skirts	3.28	14.75	39.34	19.67	16.39	6.56	46.90
Trousers/shorts	2.67	33.33	38.67	5.33	17.33	2.67	45.52
T-shirts	0.00	15.94	21.74	14.49	40.58	7.25	48.97
Jackets/coats	14.29	30.61	24.49	4.08	12.24	14.29	32.41
Any Other	0.00	4.35	43.48	4.35	21.74	26.09	16.55

Exhibit 4.6: Productivity of apparel firms (pieces/machine/day)

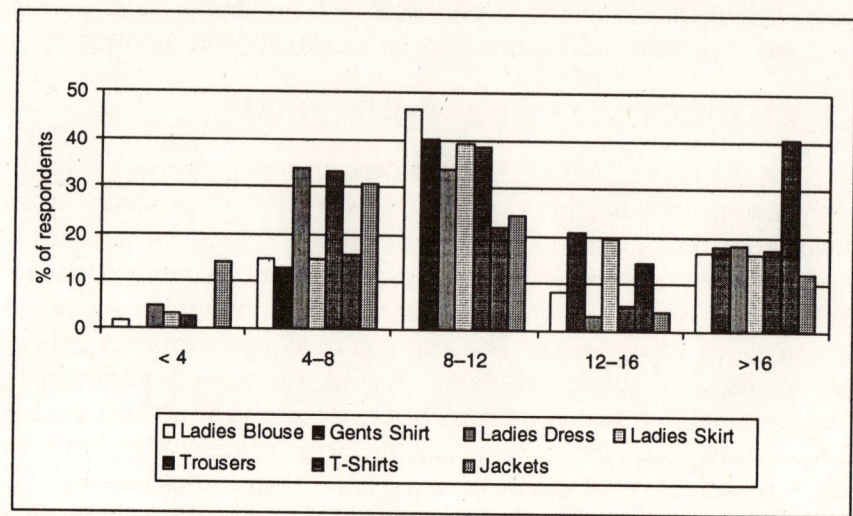

on competitiveness of Indian apparel industry. The reasons can be linked to low investment in technology and lesser number of machines in Indian industry causing technological backwardness of Indian apparel industry leading to low productivity. The lower productivity acts as deterrent for large size orders and hence affects performance and image in world market.

4.4   REASONS OF EXPORTING/SOURCING WOVEN VS. KNITTED APPAREL

There is an increasing demand of knitted apparels in world trade while Indian industry still continues to put more emphasis on woven apparels. The analysis of perceptions of apparel exporters and buying houses is undertaken to understand it.

The level of importance for various parameters has been drawn on a scale of 1-5 (1 for least important to 5 most important). The weights are assigned and weighted average scores are shown in Table 4.10 & 4.11 and Exhibit 4.7 and 4.8.

4.4.1 Response of Apparel Exporters

The sample consisted of 37.14 percent of the respondents' only dealing with woven apparels while 28.57 percent of them are dealing with only knitted apparels. Table 4.10 & Exhibit 4.7 show that the analysis of survey of apparel exporters indicates that woven apparels are perceived to have better market demand, margin, brand image, technological competence besides quota availability and favorable government policies. The knitted apparels are perceived to have better raw material availability, lesser investment requirement and availability of finishing facility leading to decision for manufacturing / exporting / sourcing of woven or knitted apparels.

Table 4.10: Reasons for exporting woven/knitted apparel

Parameters	Woven apparels (Average score)	Knitted apparels (Average score)
Market demand	4.20	3.75
Margin	3.75	3.07
Brand image	3.49	3.30
Technology competence	3.87	3.15
Quota availability	3.84	3.54
Raw material availability	3.65	3.90
Investment required	3.49	3.59
Favorable govt. policy	3.33	3.15
Availability of finishing facility	3.57	2.85

Exhibit 4.7: Level of importance for exporting woven/knitted apparel

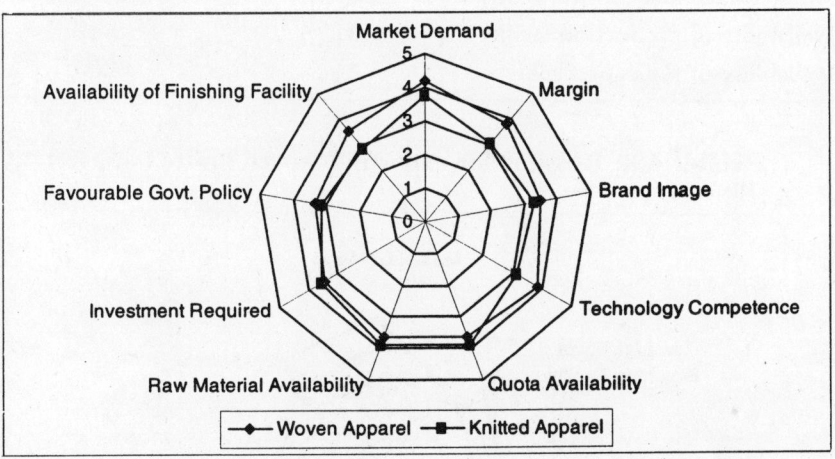

The decision to export woven or knitted apparel is based upon the perception that woven apparels are having better market demand, margin and importance of brand image, quota-availability and favorable government policy beside having technological competence while knitted apparels are favoured for exporting or sourcing due to better raw-material availability of finishing facility etc.

4.4.2 Response of Buying Houses

The sample profile of buying houses consisted of around 71 percent of the respondents primarily sourcing woven apparels and the rest were having more percentage of knitted apparels in their sourcing of apparels from India. Table 4.11 & Exhibit 4.8 show the reasons for exporting/sourcing woven vs. knitted apparels and indicate that better margin, brand image, technology competence, quota availability, availability of the production facility and availability of the finishing facility for woven apparels are the key reasons for opting for sourcing of more of woven apparels in comparison to knitted ones while the knitted apparels are having better market demand. The demand in terms of volume seems to be more for the knitted apparels while in all other parameters woven

Table 4.11: Reasons for exporting woven/knitted apparel

Parameters	Woven apparels (Average score)	Knitted apparels (Average score)
Market demand	4.30	4.42
Margin	3.71	3.38
Brand image	3.79	3.74
Technology competence	3.75	3.53
Quota availability	4.08	3.39
Availability of production facility	3.81	3.47
Availability of finishing facility	3.24	3.00

Exhibit 4.8: Reasons for exporting woven/knitted apparel

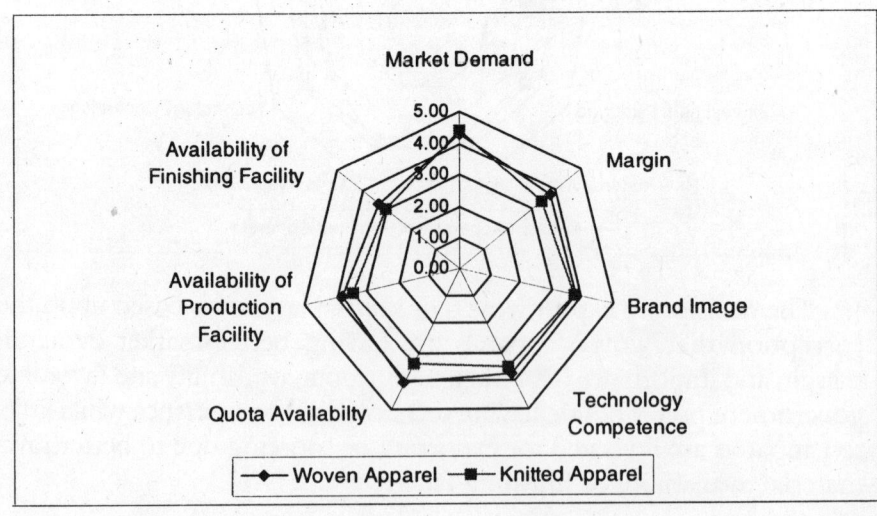

apparel has got higher weighted average score indicating favorable business terms.

The analysis of reasons for sourcing knitted apparel indicate better market demand while for all other parameters i.e. margin, technology competence, quota availability, availability of production facility, availability of finishing facility the woven apparels are being preferred.

As discussed earlier, the world trade is shifting towards knitted apparels while India is still concentrating on woven apparels, primarily due to better margins and more demand. There is a good growth in volume terms while value terms it still lags behind woven apparels, which is leading exporters to concentrate on woven apparels. The phasing out of quotas shall help in increasing exports of knits from India as earlier despite of increasing market demand due to constraints in availability of quotas, woven apparel are preferred by exporters.

4.5 AVAILABILITY OF FABRIC FOR APPAREL FROM VARIOUS SOURCES

The availability of raw material i.e. fabric is one of the key parameters determining competitiveness of apparel industry. The response of apparel exporters and buying houses has been collected to understand competitiveness of Indian fabric industry with a perspective of sourcing raw material. The sample of survey of apparel exporters consisted of 54.20 percent of respondents importing the fabric for their units. Table 4.12 indicates that respondents consuming more than 50 percent of imported fabric in their fabric consumption constitute only 22.14 percent of total sample of study.

Table 4.12: Pattern of consumption of imported fabric

Consumption of Imported Fabric in Company	*percent of Respondent*
100%	3.82
>50%	18.32
< 50%	32.06
Nil	45.80

The various key fabrics being sourced for manufacturing of apparel are being studied separately so as to analyze the competitive position of India in various fabrics as the raw material for apparel in world fabric market. The response of apparel exporters and buying houses against their preferred source (country) is depicted in tables exhibit respectively for various fabrics.

4.5.1 Preference for sourcing 100 percent cotton fabric

The analysis reflects that India followed by China is the key sourcing destinations for the 100 percent cotton fabric for apparel exporters as well as buying houses. The other countries i.e. South Korea, Taiwan etc. are being preferred by a few respondents only (Table 4.13 & Exhibit 4.9).

Table 4.13: Preference for sourcing 100 percent cotton fabric

	India	China	S.Korea	Taiwan	Malaysia	Japan	Italy
Apparel exporters	67.23	18.51	3.83	4.68	1.49	0.21	2.77
Buying houses	60.84	26.51	4.22	4.22	0	0	0

Note: All the figures indicated percentage of respondents

Exhibit 4.9: Preference for sourcing 100 percent cotton fabric

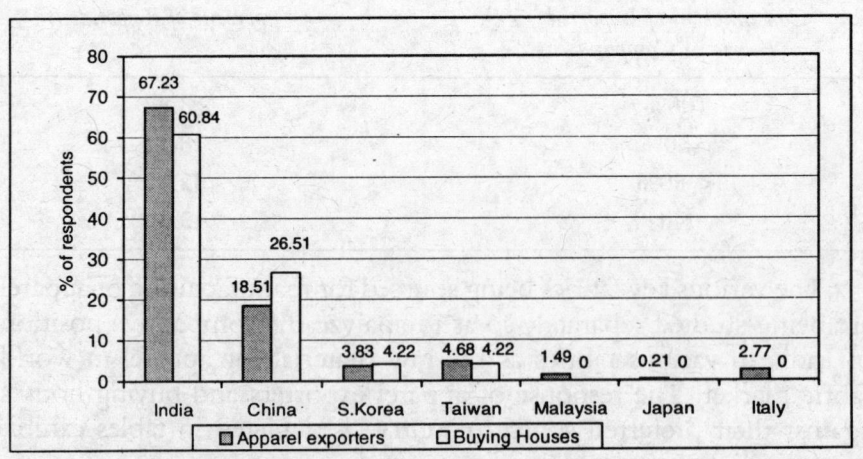

4.5.2 Preference for sourcing 100 percent MMF

China is the most preferred source of 100 percent man-made fabric followed by India, Taiwan and South Korea (Table 4.14 & Exhibit 4.10). The response of buying houses indicate preference of China, Taiwan followed by India while the response of apparel exporters indicate China followed by India and Taiwan as preferred source for MMF.

Table 4.14: Preference for sourcing 100 percent MMF

	India	China	S.Korea	Taiwan	Malaysia	Japan	Italy
Apparel exporters	27.72	32.12	12.69	20.73	1.04	1.81	0.78
Buying houses	15.72	47.8	15.09	18.87	0	0	0

Note: All the figures indicated percentage of respondents

Exhibit 4.10: Preference for sourcing 100 percent MMF

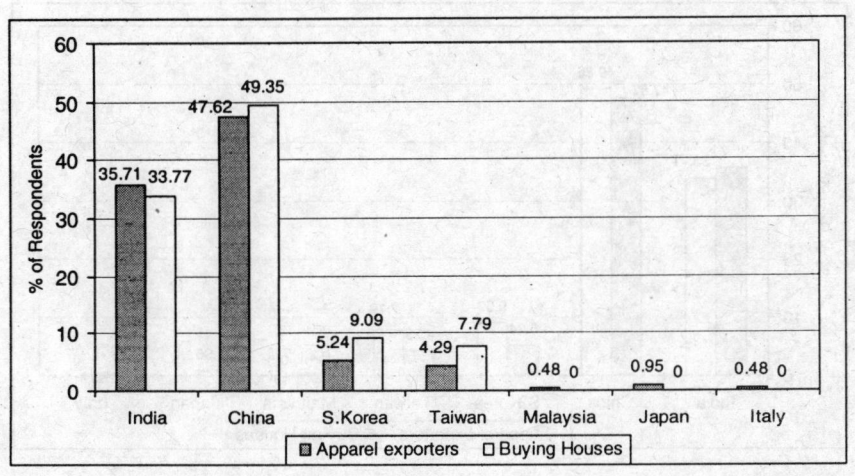

4.5.3 Preference for sourcing silk fabric

China is the most preferred sourcing destination followed by India by apparel exporters as well as buying houses; the other sourcing destinations are South Korea, Taiwan (Table 4.15 & Exhibit 4.11).

Table 4.15: Preference for sourcing silk fabric

	India	*China*	*S. Korea*	*Taiwan*	*Malaysia*	*Japan*	*Italy*
Apparel exporters	35.71	47.62	5.24	4.29	0.48	0.95	0.48
Buying houses	33.77	49.35	9.09	7.79	0	0	0

Note: All the figures indicated percentage of respondents

Exhibit 4.11: Preference for sourcing silk fabric

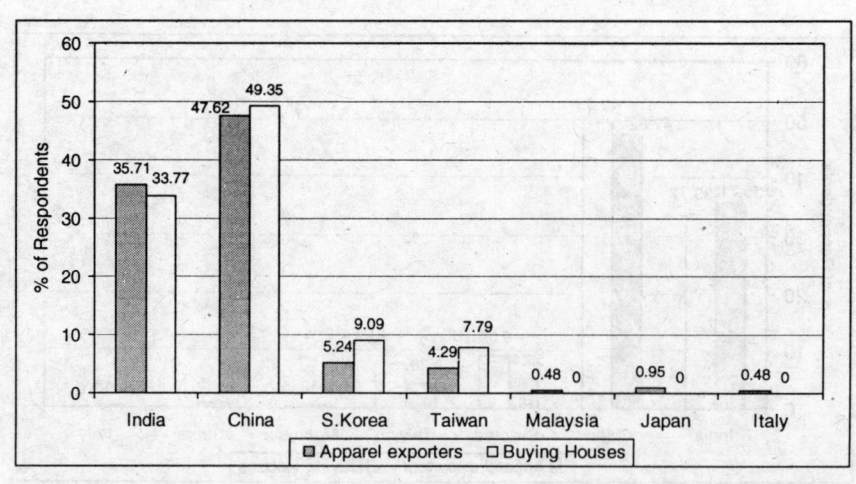

4.5.4 Preference for sourcing 100 percent wool fabric

India is the most preferred destination for sourcing 100 percent wool fabric; the other countries i.e. Italy and China are also being preferred/ recommended by the companies. It is important to note that India is being preferred by apparel exporters while buying houses prefer China and Italy more than the apparel exporters (Table 4.16 & Exhibit 4.12).

Table 4.16: Preference for sourcing 100 percent wool fabric

	India	China	S.Korea	Taiwan	Malaysia	Japan	Italy
Apparel exporters	52.9	9.68	7.1	4.52	0.65	2.58	14.84
Buying houses	36.36	33.77	1.3	0	0	0	22.08

Note: All the figures indicated percentage of respondents

Exhibit 4.12: Preference for sourcing 100 percent wool fabric

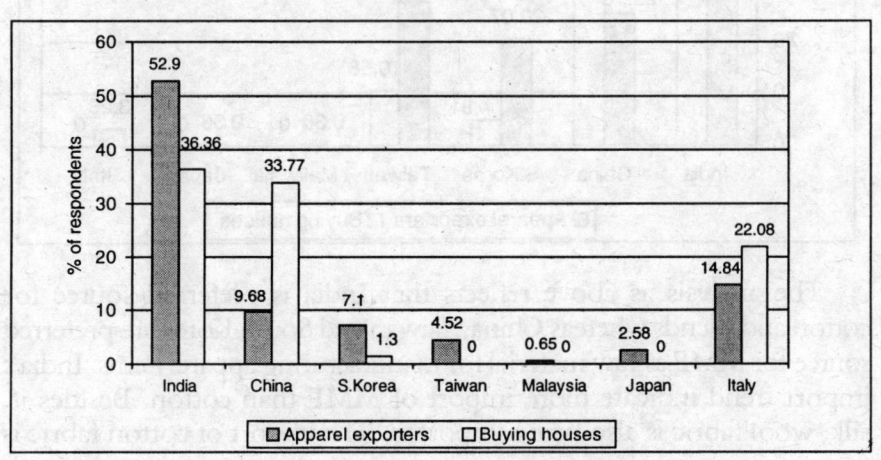

4.5.5 Preference for sourcing blended fabric

India followed by Taiwan, China and South Korea are the preferred sources of blended fabric for apparel exporters while buying houses prefer to source from China followed by India. There is strong gap between opinion of apparel exporters and buying houses regarding sourcing of blends (Table 4.17 & Exhibit 4.13). It indicates wherever the decision is with buying houses blended fabric sourced from China is preferred while exporters prefer to source from India. The blends being preferred include polyester-cotton, polyester-viscose, CVC, lycra cotton, poly-wool, cotton wool, cotton nylon and nylon-cotton-polyester etc.

Table 4.17: Preference for sourcing blended fabric

	India	China	S.Korea	Taiwan	Malaysia	Japan	Italy
Apparel exporters	65.69	25.18	20.07	27.37	0.36	0.36	3.28
Buying houses	33.65	49.04	4.81	10.58	0	0	0

Note: All the figures indicated percentage of respondents

Exhibit 4.13: Preference for sourcing blended fabric

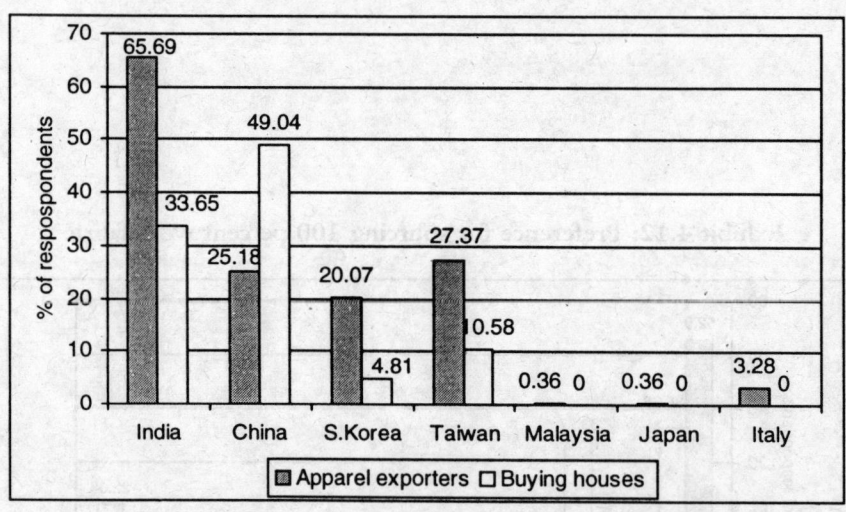

The analysis as above reflects that India is preferred source for cotton and Blends whereas China, Taiwan and South Korea are preferred source for MMF as raw material for manufacturing apparel items. India's import trend indicate more import of MMF than cotton. Besides it, silk, wool fabric is also being imported. The import of cotton fabric is also significant but not much growth is there. The cotton fabric may be also being imported due to better quality, price, superior finish etc. despite of availability in local market. This shall affect competitiveness of Indian textile industry and so is being studied separately in sixth chapter. The perception of exporters and buying houses towards various fabric sources against different parameters are studied in Chapter 5 in detail.

4.6 SUMMARY

It can be summarized in the end of this chapter that India primarily caters to lower-end of apparel market in basic apparel categories. The productivity of Indian firm is rather poor in comparison to world. Both of these facts give the image of low-end operator to India in world apparel market and cause lesser sourcing of high value item from India.

The statistical (factor) analysis also indicates that raw material & image function and market & cost functions are key reasons for lesser focus on sourcing of high value items from India. The image of India as basic producer of basic items and poor availability of requisite raw material required for high value items alongwith difficulty in locating market contributes in unfavorable cost competitiveness of Indian products in high value item market in world apparel trade resulting in lesser focus on exporting and sourcing of high value items from India.

The apparel firms are less cost competitive due to lower productivity, low R&D, technology, non-availability of finishing facilities causing for not being equipped to cater to higher end of the market with relatively more margin. The unfavorable cost-competitiveness is also caused due to non-availability of raw-material particularly MMF which is being increasingly imported form China, Taiwan, and Koreas as reflected in primary as well as secondary research during the study. India's produces more of cotton while the demand in international market is of MMF, which indicate requirement of fabric of MMF composition for apparel exporters and leads to import of fabric by apparel exporters. There is need to realign the fiber mix of India's textile production to stay competitive in world market A further analysis of competitiveness of textile industry has been discussed in the following Chapter 5 so as to understand the factors from textile Industry affecting competitiveness of Indian textile and apparel firms in world trade.

chapter **5**

Competitiveness of Indian Textile (fabric) Industry: Study of Perceptions

The Indian textile and apparel trade is facing a tough competition due to liberalization of trade under WTO. Besides it, there are opportunities for expansion of trade in post-2004 period. The Indian textile sector is known in world market for its distinctive offerings. The textile (fabric) sector caters to domestic market of apparel as well as of apparel exporters besides export of fabric. To understand the competitiveness of Indian fabric sector the inputs were taken from apparel exporters and buying houses along with fabric manufacturers so as to have a unified picture of Indian fabric industry in quota free period. The response has been categorized under various sub-themes and is discussed in detail.

The various sub-themes are:

1. Fabric sources for apparel industry
2. Composition of fabric manufactured
3. Reasons for not entering into fabric export market
4. Lesser focus on manufacturing of high-value fabric.

5.1 FABRIC SOURCES FOR APPAREL INDUSTRY

The sources (countries) of fabric for apparel play an important role in determining the cost of apparel, lead-time and in turn competitiveness of the firm. The responses were taken from apparel exporters and buying houses so as to understand the preferred sources of key fabrics being used in the trade. The preference towards various fabric sources while making sourcing decisions has been studied in detail.

The apparel exporters and / or buying houses have a choice to source the fabric from organized, unorganized or imported source of fabric. The response of apparel exporters as well as buying houses has been taken against various parameters playing an important role in sourcing decision for the fabrics; the weighted average scores are ranging from 1-5 (1 being least important; 5 being most important). The weights were assigned and weighted average score are shown in Table & Exhibit 5.1–5.2 and are also depicted with help of radar diagram. The response of apparel exporters and buying houses indicating the preference for sources of fabric has been compared for all the parameters individually (Table & Exhibit 5.3–5.14) so as to understand difference in perception of apparel exporters and buying houses for suitability of various fabric sources against key parameters.

5.1.1 Response of Apparel Exporters

Table & Exhibit 5.1 shows that the response of apparel exporters indicate the preference for imported fabric on account of lower price, availability of wider width of fabric, consistency in quality, required quality of finishing, consistency in lot/roll quantity, lot / roll quantity availability and the availability of counts and construction as per requirement. The fabric from organized sector is perceived to be preferred while sourcing fabric with target of high order quantity and minimum lead-time. Required quality of processing and physical properties is also perceived to be superior in organized sector fabric as indicated by apparel exporters. The fabric from un-organized sector is only perceived to be preferred on account of low sampling cost.

Table 5.1: Level of preference of various fabric sources

Parameters	Organised	Unorganised	Imported
High order quantity	4.13	2.71	3.93
Lower price	3.92	3.54	3.96
Wider width of fabric	4.00	2.68	4.07
Minimum lead time	4.05	3.34	3.85
Consistency in quality	4.16	2.78	4.52
Required quality of processing	4.27	2.88	4.18
Required quality of finishing	4.14	2.85	4.23
Consistency in lot/roll quantity	3.97	2.61	4.61
Lot/roll quantity availability	3.88	2.63	4.16
Count and construction availability	3.96	3.00	4.23
Low sampling cost	3.48	3.79	2.71
Required physical properties	4.12	3.06	4.07

Exhibit 5.1: Level of preference of various fabric sources

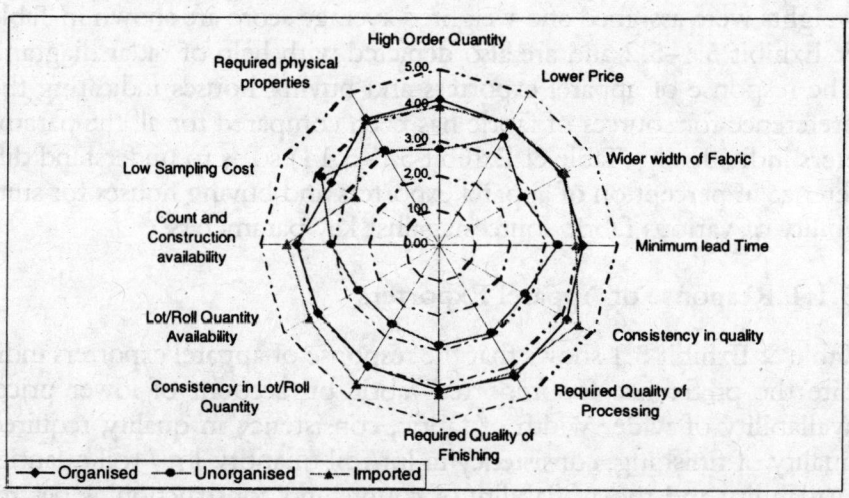

5.1.2 Response of Buying Houses

The response of buying houses indicate that imported fabric is perceived to be superior to other source of fabric i.e. unorganized & organized mills, while considering various parameters for making sourcing decision regarding fabrics for apparel items being sourced form India. Table 5.2 & Exhibit 5.2 shows that the fabric from unorganized sector is perceived to be better than fabric from organized sector and preferred while considering lower price and low sampling cost as decision making criteria for sourcing decision.

Table 5.2: Level of preference of various fabric sources

Parameters	Organised	Unorganised	Imported
High order quantity	3.90	1.81	4.68
Lower price	3.26	3.43	4.35
Wider width of fabric	3.52	1.70	4.70
Minimum lead time	3.45	2.89	4.03
Consistency in quality	3.66	1.97	4.73
Required quality of processing	3.36	2.07	4.69
Required quality of finishing	3.44	1.50	4.72
Consistency in lot	3.65	2.00	4.71
Lot quantity available	3.72	2.44	4.55
Count and construction availability	3.84	2.56	4.61
Low sampling cost	3.13	3.31	3.43
Required physical properties	3.61	1.68	4.59

Exhibit 5.2: Level of preference of various fabric sources

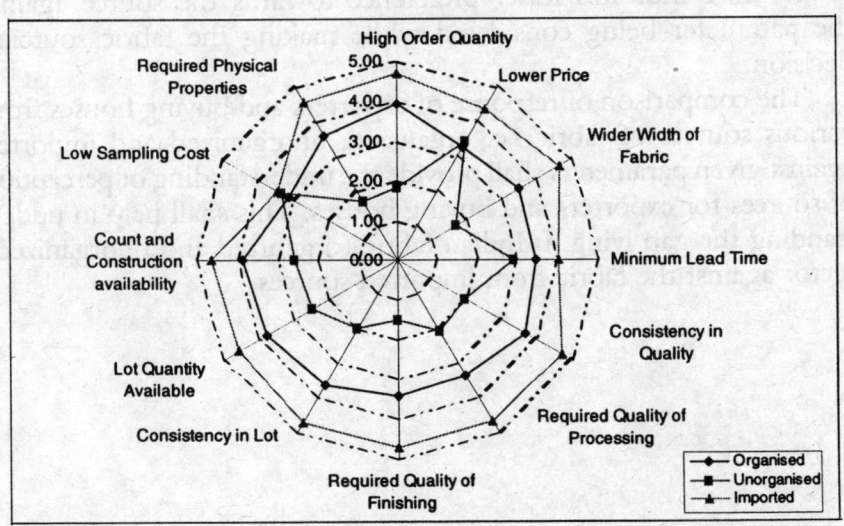

In all other respects, response of buying houses indicate that the performance of fabric from organized sector is preferable than fabric form unorganized sector. However, there is an overall perception of best performance of imported fabric. Since in many cases buying houses recommend or play an important role in selection of fabric sources for raw material for apparel exporter, their view are critical and to be considered while planning action plan for making Indian fabric industry competitive in world in post-MFA period.

5.1.3 Comparative analysis of organized, un-organized and imported source of fabric

The comparative analysis of sources of fabric i.e. organized, unorganized and imported has been done based upon the response from apparel exporters and buying houses. The responses of exporters and buying houses are shown individually against each of the parameters considered important while making sourcing decision. Various sources of procurement of fabric by apparel exporters or buying houses are compared against given parameters. The lesser score indicate need for improvement against given parameter for changing the perception or in turn increasing competitiveness of textile trade. The response have been collected for the parameters important in making the sourcing decision. The weighted average scores are on a scale of 1-5 (1 being least important & 5 being most important). Weights were assigned and average scores are shown in table as well as represented in radar diagram. The higher score indicates preference towards the source while lower score indicates lesser preference towards the source against the parameter being considered while making the fabric sourcing decision.

The comparison of response of exporters and buying houses from various sources of fabric i.e. organized, unorganized and imported against given parameters shall provide the understanding of perception of sources for exporters and buying houses. This shall help in understanding the gap lying in Indian fabric (organized and unorganized) sector against the fabric from imported sources.

5.1.3.1 Perception towards high order quantity

While sourcing the fabric with high order quantity; apparel exporters prefer the fabric from organized source followed by imported fabric while buying houses have strong preference towards imported fabric followed by organized source. The fabric from unorganized sector is having least preference from both sets of the respondents (Table & Exhibit 5.3).

Table 5.3: Perception of fabric source against high order quantity

Source Response	Organised	Unorganised	Imported
Apparel exporters	4.13	2.71	3.93
Buying houses	3.90	1.81	4.68

Note: Score on a scale of 1-5

Exhibit 5.3: Perception of fabric source against high order quantity

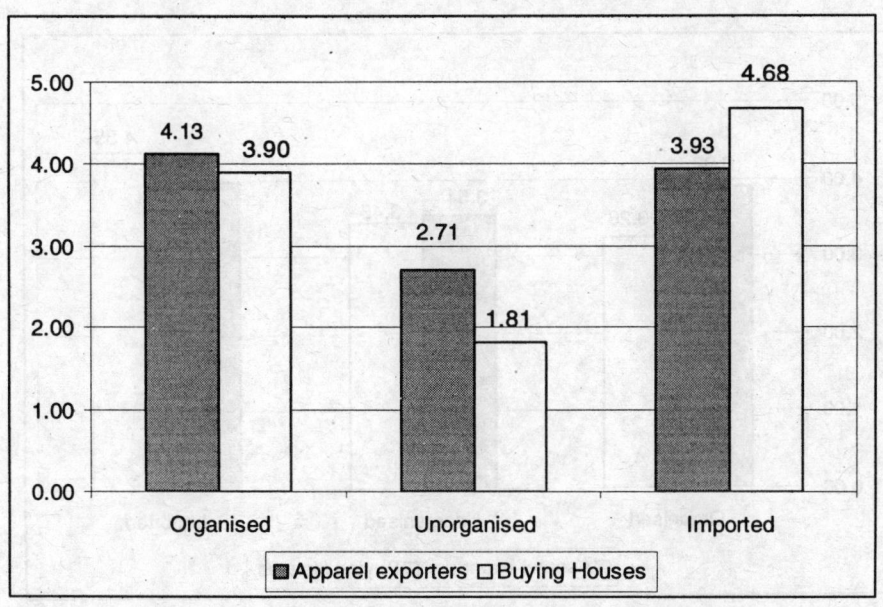

5.1.3.2 Perception of fabric source against lower price

The response of apparel exporters as well as buying houses indicate that price being criteria of sourcing of fabric for apparel firms; imported fabric shall be preferred followed by fabric from organized and unorganized sector. (Table and Exhibit 5.4).

Table 5.4: Perception of fabric source against lower price

Source Response	Organised	Unorganised	Imported
Apparel exporters	3.92	3.54	3.96
Buying houses	3.26	3.43	4.35

Note: Score on a scale of 1-5

Exhibit 5.4: Perception of fabric source against lower price

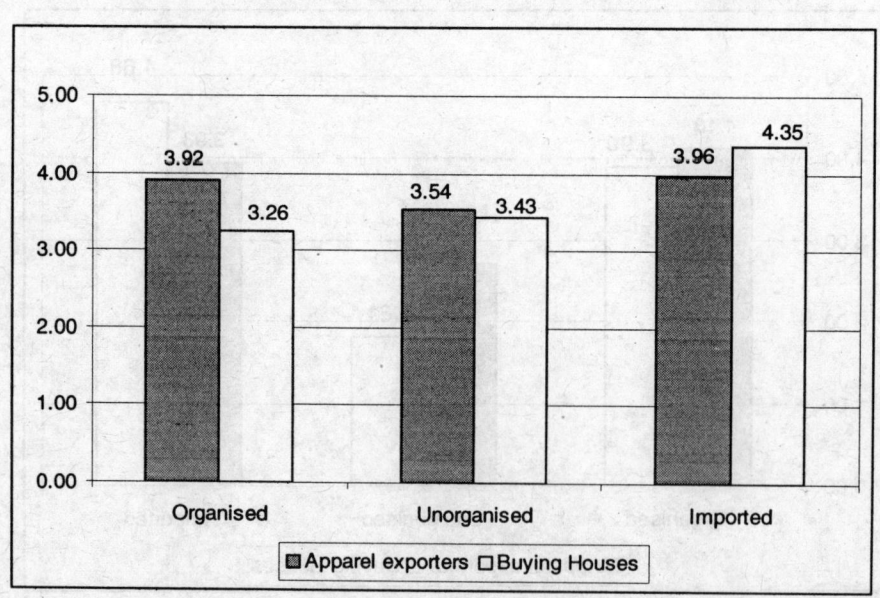

5.1.3.3 Perception of fabric source against wider width of fabric

The imported fabric is perceived to provide wider width options followed by fabric from organized sector. The fabric from unorganized sector is perceived to provide lesser width fabric (Table & Exhibit 5.5). Wider width of fabric being the criteria for sourcing decision, imported fabric followed by organized sector will be preferred for sourcing.

Table 5.5: Perception of fabric source against wider width of fabric

Response \ Source	Organised	Unorganised	Imported
Apparel exporters	4.00	2.68	4.07
Buying houses	3.52	1.70	4.70

Note: Score on a scale of 1-5

Exhibit 5.5: Perception of fabric source against wider width of fabric

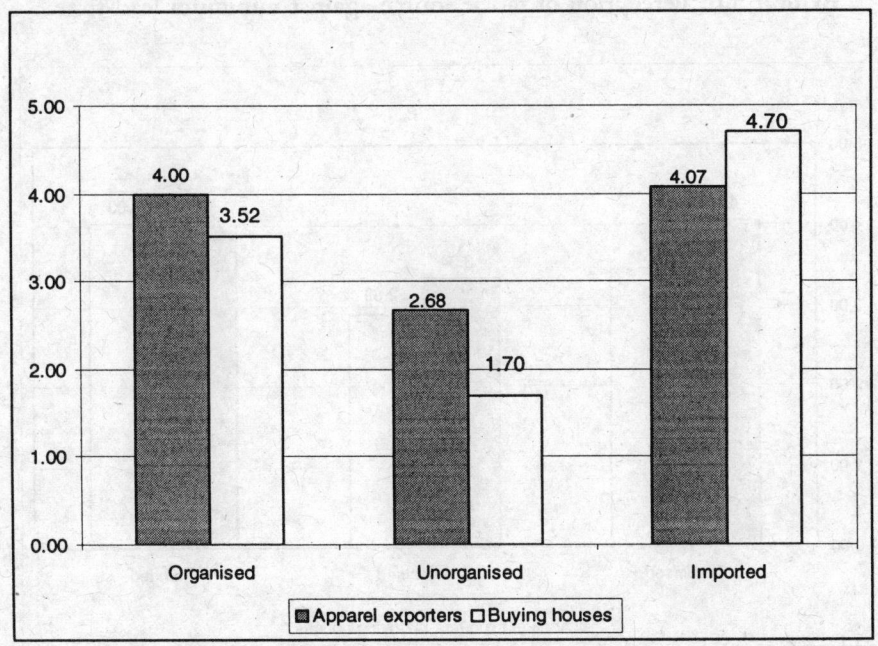

5.1.3.4 Perception of fabric source against minimum lead-time

Lead time being the criteria of comparison, imported fabric followed by fabric from organized sector is preferred by buying houses while the response of apparel exporters indicates the preference for fabric from organized sector followed by imported fabric while considering this parameter (Table & Exhibit 5.6).

Table 5.6: Perception of fabric source against minimum lead-time

Response \ Source	Organised	Unorganised	Imported
Apparel exporters	4.05	3.34	3.85
Buying houses	3.45	2.89	4.03

Note: Score on a scale of 1-5

Exhibit 5.6: Perception of fabric source against minimum lead-time

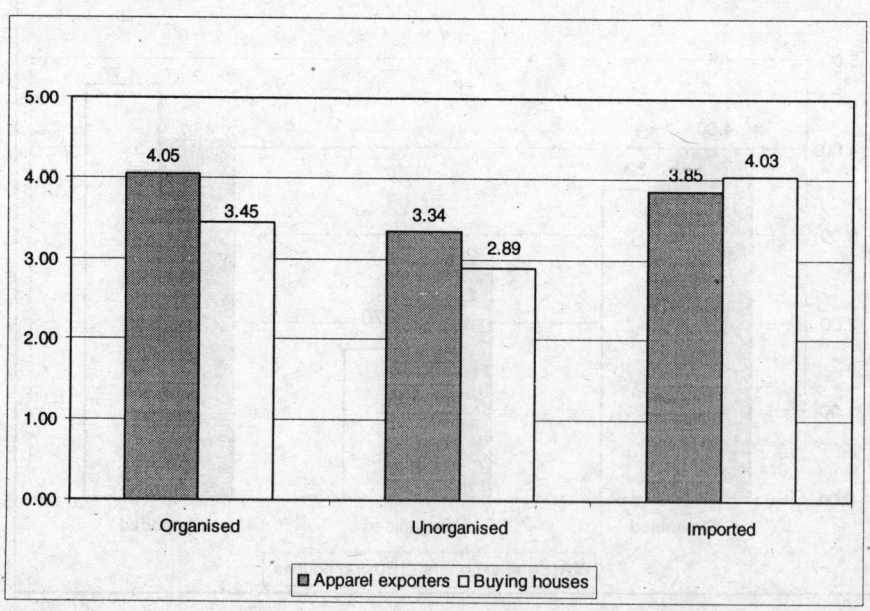

5.1.3.5 Perception of fabric source against consistency in quality

The imported fabric is perceived to have more consistency in quality in comparison to fabric from organized sector and unorganized sector as shown in Table & Exhibit 5.7. The survey of apparel exporters as well buying houses indicates the similar findings with respect to consistency in quality. The fabric from un-organized sector is perceived to be having least perceived consistency in quality.

Table 5.7: Perception of fabric source against consistency in quality

Response \ Source	Organised	Unorganised	Imported
Apparel exporters	4.16	2.78	4.52
Buying houses	3.66	1.97	4.73

Note: Score on a scale of 1-5

Exhibit 5.7: Perception of fabric source against consistency in quality

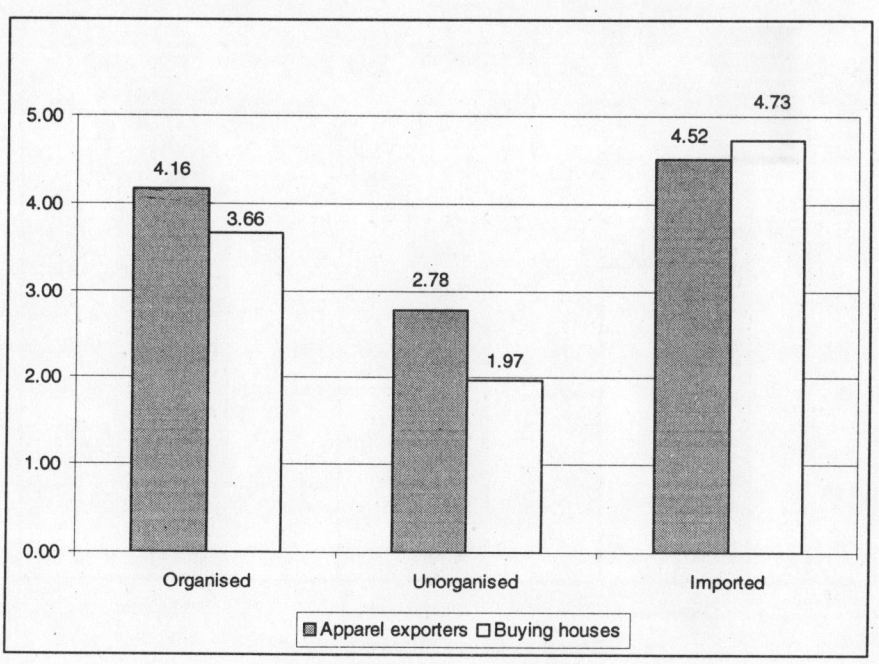

5.1.3.6 Perception of fabric source against required quality of processing

Buying houses perceive imported fabric as superior in required quality of processing while apparel exporters indicate preference for fabric from organized sector followed by imported source of fabric (Table & Exhibit 5.8). The fabric from unorganized sector is perceived to be rather poor in delivering required quality of processing and hence least preferred.

Table 5.8: Perception of fabric source against required quality of processing

Response \ Source	Organised	Unorganised	Imported
Apparel exporters	4.27	2.88	4.18
Buying houses	3.36	2.07	4.69

Note: Score on a scale of 1-5

Exhibit 5.8: Perception of fabric source against required quality of processing

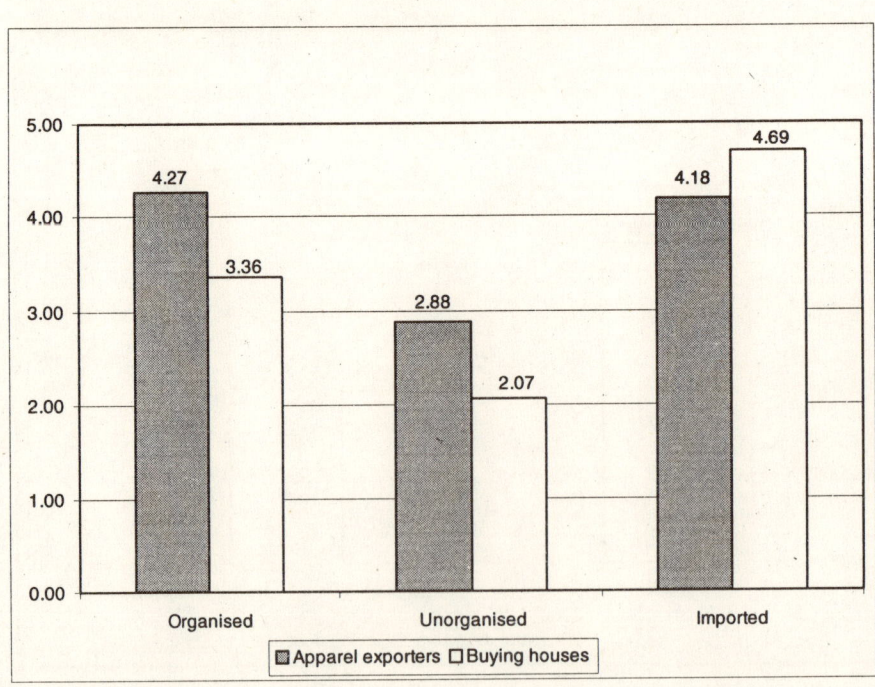

5.1.3.7 Perception of fabric source against required quality of finishing

Imported fabric is perceived to have better quality of finishing in comparison to other sources of fabric. The fabric from unorganized sector is having least perceived value against this parameter (Table & Exhibit 5.9). The imported fabric is preferred by apparel exporters as well as buying houses while considering quality of finishing as sourcing criteria.

Table 5.9: Perception of fabric source against required quality of finishing

Response \ Source	Organised	Unorganised	Imported
Apparel exporters	4.14	2.85	4.23
Buying houses	3.44	1.50	4.72

Note: Score on a scale of 1-5

Exhibit 5.9: Perception of fabric source against required quality of finishing

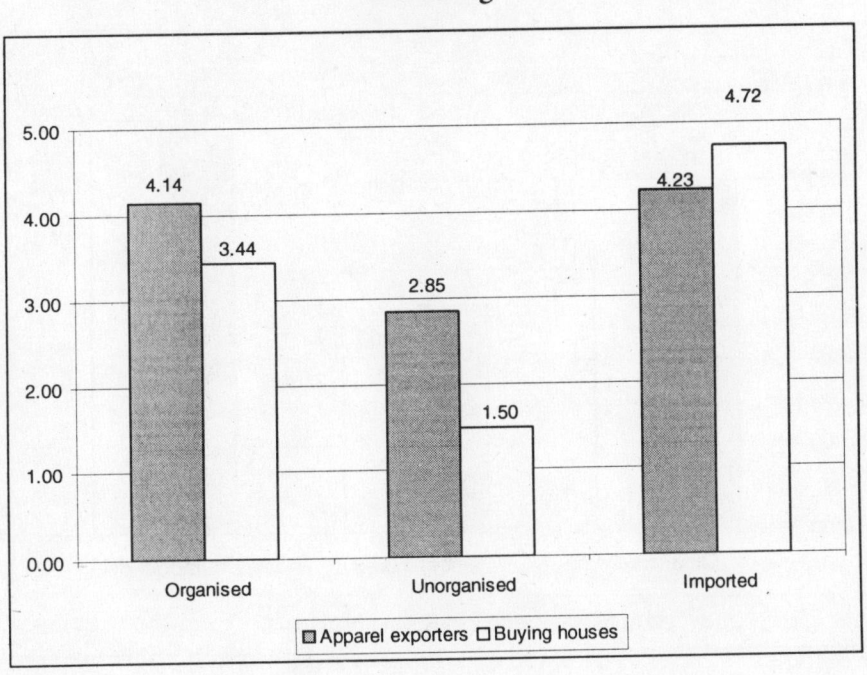

5.1.3.8 Perception of fabric source against consistency in lot

The responses of apparel exporters as well as buying houses indicate preference towards imported fabric while considering consistency in lot of the fabric. Here too, the fabric from unorganized sector is having least amount of preference (Table & Exhibit 5.10). Apparel exporters as well as buying houses prefer the imported fabric on account of consistency in lot.

Table 5.10: Perception of fabric source against consistency in lot

Source Response	Organised	Unorganised	Imported
Apparel exporters	3.97	2.61	4.61
Buying houses	3.65	2.00	4.71

Note: Score on a scale of 1-5

Exhibit 5.10: Perception of fabric source against consistency in lot

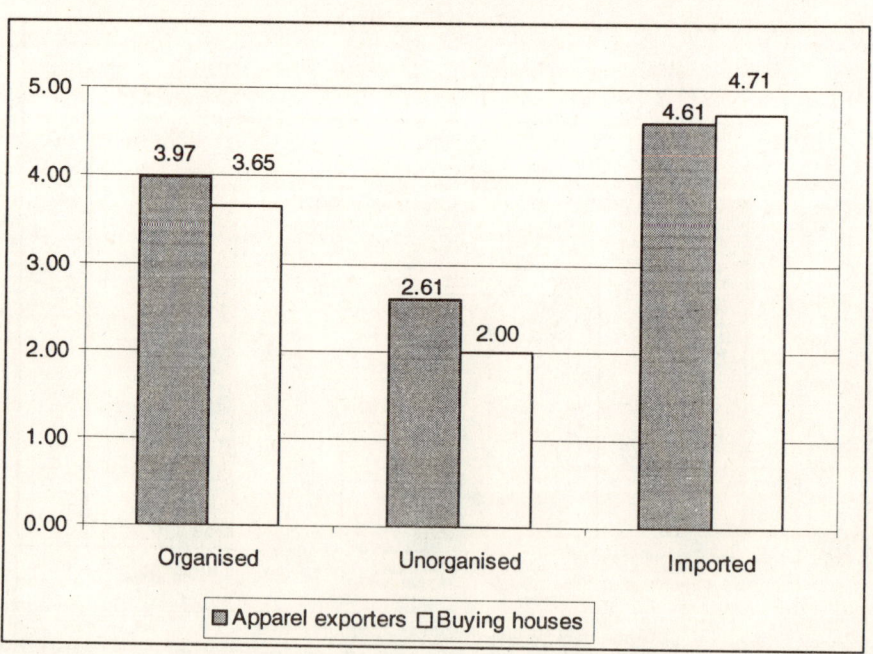

5.1.3.9 Perception of fabric source against lot/roll quantity available

Imported fabric followed by fabric from organized sector is perceived to have required lot roll quantity/ availability indicating the preference for a specific lot quantity (Table & Exhibit 5.11). The availability of lot/roll quantity in fabric from unorganized sector is rather poor.

Table 5.11: Perception of fabric source against lot/roll quantity available

Response \ Source	Organised	Unorganised	Imported
Apparel exporters	3.88	2.63	4.16
Buying houses	3.72	2.44	4.55

Note: Score on a scale of 1-5

Exhibit 5.11: Perception of fabric source against lot/roll quantity available

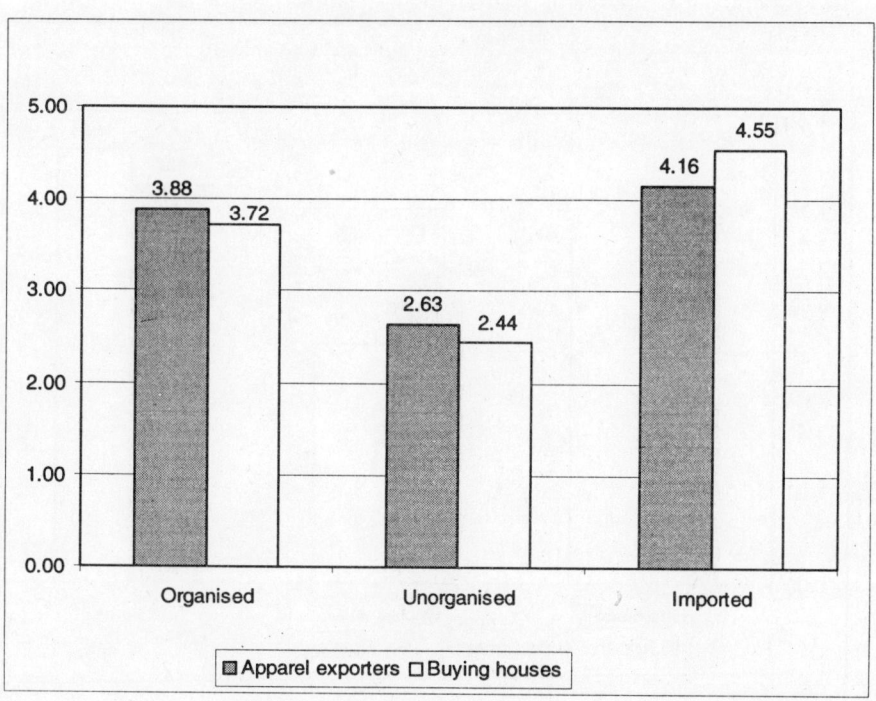

5.1.3.10 Perception of fabric source against count and construction availability

Imported fabric is perceived to have more availability of count and construction of fabrics followed by the fabric from organized sector (Table & Exhibit 5.12). The fabric from unorganized sources is perceived to have less preference on account of possibility of lesser variations in terms of count and construction availability in fabric.

Table 5.12: Perception of fabric source against count and construction availability

Response / Source	Organised	Unorganised	Imported
Apparel exporters	3.96	3.00	4.23
Buying houses	3.84	2.56	4.61

Note: Score on a scale of 1-5

Exhibit 5.12: Perception of fabric source against count and construction availability

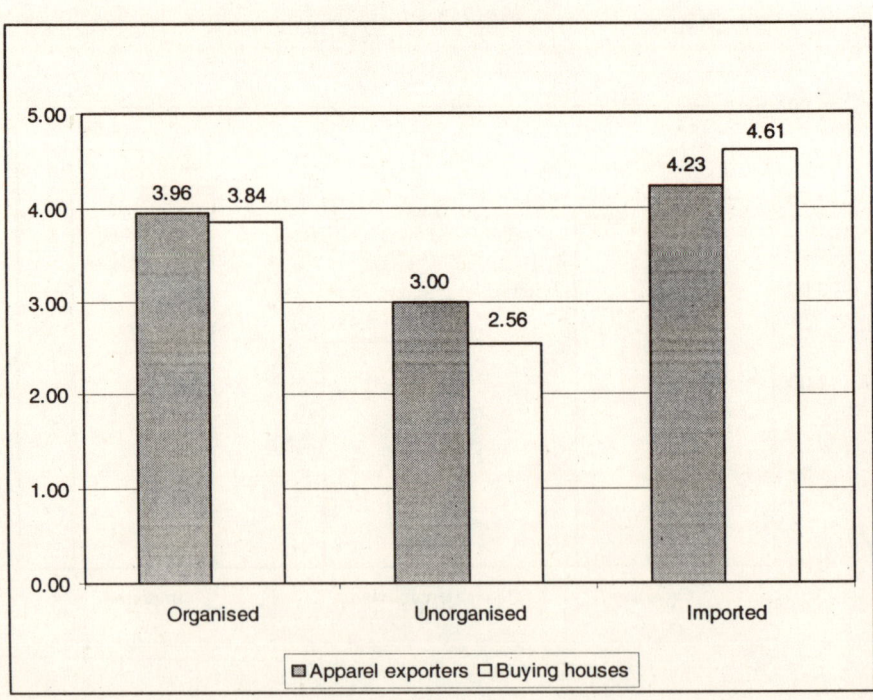

5.1.3.11 Perception of fabric source against low sampling cost

Apparel exporters indicate unorganized sector to have low sampling cost of fabric followed by fabric from organized sector. The buying houses have indicated preference for imported fabric followed by unorganized source of fabric against this parameter (Table & Exhibit 5.13).

Table 5.13: Perception of fabric source against low sampling cost

Response / Source	Organised	Unorganised	Imported
Apparel exporters	3.48	3.79	2.71
Buying houses	3.13	3.31	3.43

Note: Score on a scale of 1-5

Exhibit 5.13: Perception of fabric source against low sampling cost

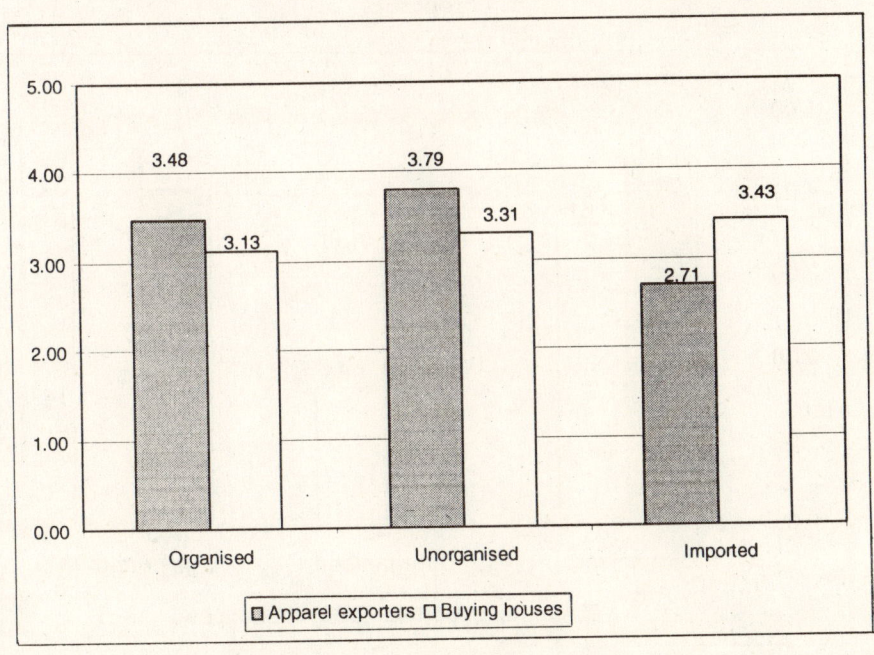

5.1.3.12 Perception of fabric source against required physical properties

Apparel exporters indicate preference for fabric from organized sector closely followed by imported fabric while considering required physical properties as the parameter for sourcing the fabric. The buying houses strongly indicate their preference towards the imported fabric. (Table & Exhibit 5.14)

Table 5.14: Perception of fabric source against required physical properties

Response \ Source	Organised	Unorganised	Imported
Apparel exporters	4.12	3.06	4.07
Buying houses	3.61	1.68	4.59

Note: Score on a scale of 1-5

Exhibit 5.14: Perception of fabric source against required physical properties

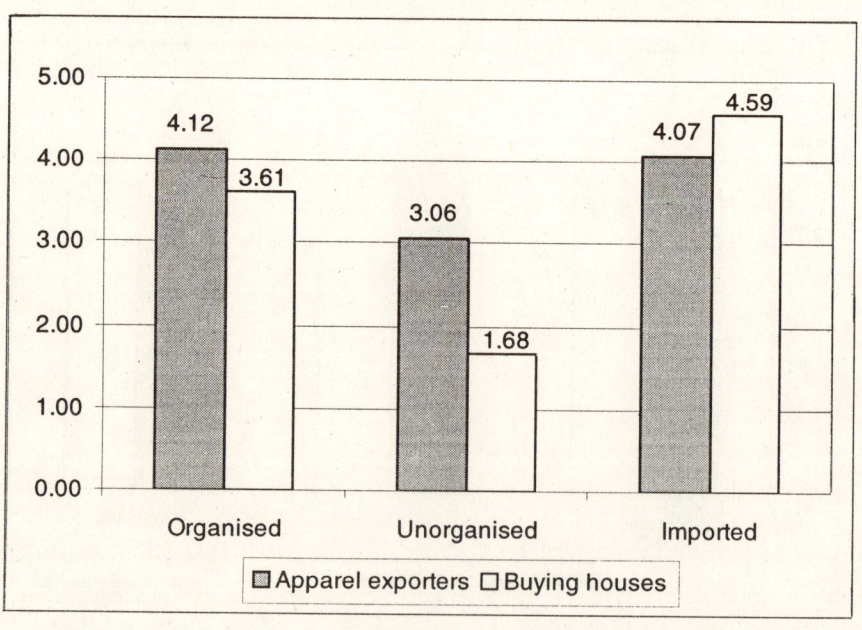

5.1.4 Statistical Analysis of Comparison of Various Sources of Fabric

Factor analysis using Principal Component Analysis (PCA) was carried out on the data collected, in order to bring out the salient features of the apparel exporters perception about various sources of fabric as raw material for the apparel. The factor analysis has been undertaken separately for analyzing performance of various sources of fabric i.e. Imported, Unorganized and Organized source of fabric procurement for apparel exporters.

Table 5.15: Factor groups for level of preference for Imported fabric

Parameters	Group 1	Group II
High order quantity	.352	.684
Lower price	.229	.690
Wider width of fabric	.604	.593
Minimum lead time	.229	.745
Consistency in quality	.837	.190
Required quality of processing	.781	.461
Required quality of finishing	.808	.454
Consistency in lot/roll quantity	.767	.469
Lot/roll quantity availability	.749	.259
Count and construction availability	.785	-1.806E-02
Low sampling Cost	-.267	.531
Required physical properties	.717	-2.001E-02

(i) Table 5.15 indicates the factor loadings of various parameters considered while making a decision to source fabric by apparel exporters from imported source of fabric. The variables with high factor loadings in group include consistency in quality (0.837), required quality of finishing (0.808) count and construction availability (0.785), required quality of processing (0.781), consistency in lot (0.767), lot quantity available (0.749) and required physical properties (0.717). The variables with high loadings in group 2 consist of minimum lead time (0.745), lower price (0.690) and high order quantity (0.684).

The total variance attributable to the first factor group is 52.622 percent. The total variance for factor group II is 64.88 percent.

Table 5.16 indicates that imported fabric is being preferred for superior performance against various parameters related to quality functions namely consistency in quality (Factor loading 0.837), required quality of finishing (Factor loading 0.808), count and construction availability (Factor Loading 0.785), required quality of processing (Factor Loading 0.781), consistency in lot (Factor loading 0.767), lot quantity available (Factor loading 0.749) and required physical properties (Factor loading 0.717). These variables (six) are highly inter correlated and combine into factor I. Due to common nature of these variables, they are represented as "Quality functions."

Table 5.16: Level of preference for Imported fabric

Parameters	Factor loading	Factor group title
Consistency in quality	0.837	
Required quality of finishing	0.808	
Count and construction availability	0.785	
Required quality of processing	0.781	Quality functions
Consistency in lot	0.767	
Lot quantity available	0.749	
Required physical properties	0.717	

Table 5.17: Level of preference for Imported fabric

Parameters	Factor Loading	Factor Group Title
Minimum lead time	0.745	
Lower price	0.690	Price and delivery Functions
High order quantity	0.684	

The other factors important for decision of sourcing imported fabric as raw material for apparel includes minimum lead-time (Factor loading 0.745), lower price (Factor loading 0.690) and requirement of high order quantity (Factor loading 0.684) and are represented by high factor loading in factor group II (Table 5.17). The factor group II is titled as" Price and delivery Function".

(ii) Table 5.18 indicates the factor loadings of various parameters considered while making a decision to source fabric by apparel exporters from unorganized source. The parameters are shown under group I and group II alongwith the factor loadings in respective groups. The parameters with high loadings in one group are put together for further analysis and titled with factor group title.

Table 5.18: Factor groups for level of preference for Unorganized fabric source

Parameters	Group I	Group II
High order quantity	.572	.519
Lower price	.311	.843
Wider width of fabric	.727	.337
Minimum lead time	.501	.667
Consistency in quality	.835	.360
Required quality of processing	.876	.292
Required quality of finishing	.861	.311
Consistency in lot/roll quantity	.845	.277
Lot/ roll quantity availability	.820	.217
Count and construction availability	.744	.344
Low sampling cost	5.824E-02	.797
Required physical properties	.755	3.723E-02

The total variance attributable to the first factor group is 62.22 percent while the total variance for factor group II is 72.09 percent. Table 5.19 indicates that in the case of unorganized source of fabric Factor group I (Quality functions) has variables linked with quality related parameters and are highly inter correlated. Processing (Factor loading 0.876), finishing (Factor loading 0.861) and consistency in quality (Factor loading 0.835) as well as in lot (Factor loading 0.845) are the variables with relatively high factor loading and hence is important in decision making for sourcing of fabric from unorganized source.

Table 5.19: Level of preference for Unorganized fabric source

Parameters	Factor Loading	Factor group title
Required quality of processing	0.876	
Required quality of finishing	0.861	
Consistency in lot	0.845	
Consistency in quality	0.835	Quality functions
Lot quantity available	0.820	
Required physical properties	0.755	
Count and construction availability	0.744	
Wider width of fabric	0.727	

The factor group II represents Price and delivery functions. The important variables are Lower price (Factor loading 0.843), low sampling cost (Factor loading 0.797) and minimum lead times (Factor loading 0.667). These variables have a high loading on the second factor as shown in Table 5.20.

Table 5.20: Level of preference for Unorganized fabric source

Parameters	Factor loading	Factor group title
Lower price	0.843	
Low sampling cost	0.797	Price and delivery Functions
Minimum lead time	0.667	

(iii) Table 5.21 indicates the factor loadings of various parameters considered while making a decision to source fabric by apparel exporters from organized source. The parameters with high factor loadings in each of these groups are representing factors with high correlation and hence given a factor group title.

Table 5.21: Factor groups for level of preference for Organized fabric source

Parameters	Group I	Group II	Group III
High order quantity	.487	.567	4.605E-02
Lower price	.142	-7.840E-02	.871
Wider width of fabric	.341	.697	-2.461E-02
Minimum lead time	.149	.335	.722
Consistency in quality	.797	4.741E-02	.266
Required quality of processing	.856	5.707E-02	.129
Required quality of finishing	.793	.175	3.450E-02
Consistency in lot/roll quantity	.796	.315	2.385E-02
Lot/roll quantity availability	.575	.522	.113
Count and construction availability	.496	.311	.277
Low sampling cost	-.127	.656	.385
Required physical properties	.514	.235	.426

The total variance attributable to the first factor group is 42.87 percent. The total variance for factor group II is 54.71 percent and for factor group III is 63.25 percent. Table 5.22 indicates in the case of organized source of fabric the variables required quality of processing

(Factor loading 0.856), consistency in quality (Factor Loading 0.797), consistency in lot (Factor loading 0.796) and required quality of finishing (Factor loading 0.793) are highly inter correlated and combined into factor I, they have been identified as quality functions. These variables are important while making sourcing decision of fabric.

Table 5.22: Level of preference for Organized fabric source

Parameters	Factor loading	Factor group title
Required quality of processing	0.856	Quality function
Consistency in quality	0.797	
Consistency in lot	0.796	
Required quality of finishing	0.793	

The variables, lower price (Factor loading 0.871) and minimum lead-time (Factor loading 0.722) have a high factor loading on the IInd factor. They are interpreted as price and delivery function (Table 5.23).

Table 5.23: Level of preference for Organized fabric source

Parameters	Factor loading	Factor group title
Lower Price	0.871	Price and delivery function
Minimum Lead Time	0.722	

The analysis of responses for various source of fabric for apparel indicates that the criteria for sourcing of fabric are determined primarily by quality functions represented by various parameters (variables). This is followed by price and delivery function. The analysis further indicate that the key variables determining source of raw material for apparel is based upon assessment of quality of various options available to apparel exporter.

5.2 COMPOSITION OF FABRIC MANUFACTURED

5.2.1 Greige vs. Finished

The fabric manufactures are manufacturing greige as well as finished fabric. The finished fabric is known to have better price realization in international market. The response of fabric manufacturers was taken to understand the reasons for manufacturers opting for manufacturing of greige fabric and not moving towards value added processed fabric.

Table 5.24: Composition of fabric manufactured (Greige/finished)

Composition of fabric manufactured	Percentage of respondents
100% Greige	24.24
> 50% Greige rest finished	9.09
> 50% Finished rest greige	1.52
100% Finished	65.15

The target sample analysis (Table 5.24) indicates that 34.85 percent of the respondents of the study are involved in manufacturing of greige fabric while 65.15 percent of the respondents are only manufacturing finish fabric.

The fabric exporters dealing with greige fabric indicate (Table 5.25) more demand and lesser marketing effort as the key reasons for exporting greige fabric followed by the other reasons including high finishing cost, low finish quality and non-availability of technology.

Table 5.25: Reasons for exporting greige fabric

Reasons for exporting greige variety	Percentage of respondents
Low finish quality	9.09
Non-availability of technology	9.09
High finishing cost	9.09
Lesser marketing effort	27.27
More demand	27.27

High market demand of greige fabric and easiness in marketing greige fabric are the foremost important reasons for manufacturing greige fabric followed by expertise in the product and less competition in the markets, investment constraint for finishing facility (Table 5.26 & Exhibit 5.15). Lesser competition in the international market for greige fabric is also an important reason for fabric manufacturers manufacturing greige fabric although profit margins and UVR is higher for finished fabric.

Table 5.26: Reasons for manufacturing greige fabric

Reasons for manufacturing greige fabric	Percentage of respondents
High market demand	19.35
Higher UVR/profit margin	2.15
Expertise in the product	16.13
Less competition in the market	12.90
Investment constraint for finishing facility	12.90
Lack of technological know how	5.38
High finishing cost	7.53
Easy to market	19.35

Exhibit 5.15: Reasons for manufacturing greige fabric

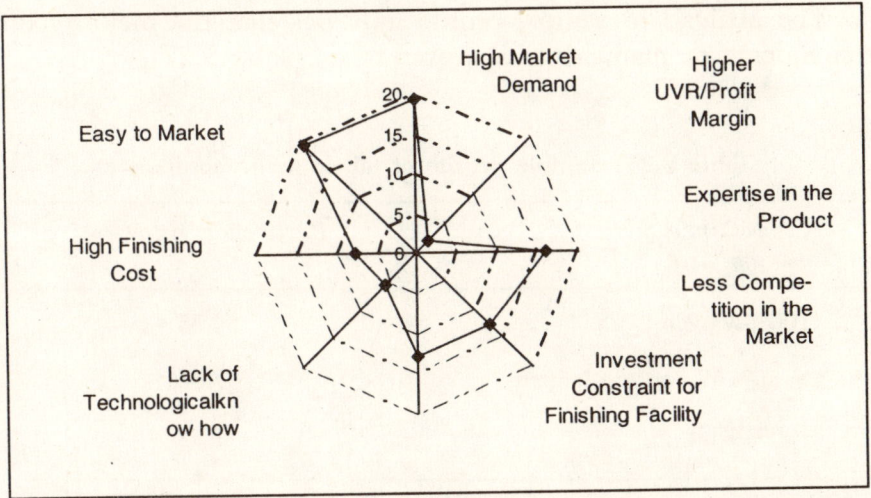

The research also indicates the more concentration of India's textile export towards greige fabric although the realizations are better or much higher in trade of processed fabric. The response of fabric manufacturers during the survey indicate the easiness to market greige fabric and market demand as the key reasons for opting for exporting or manufacturing more of the greige fabric than processed one. As already discussed the world trade in textiles is more of processed fabric due to higher UVR. The same is not the case with Indian manufacturers who are still continuing with greige fabric trade due to expertise in the product and easy approach to market, which is leading India in a less competitive position in world textile trade. The value additions are taking place in other countries where greige fabric being exported from India is processed, in a way it's opportunity loss to India. Indian manufacturers can target to upper end of fabric market or get better return provided they manufacture processed fabric.

5.2.2 Woven vs. Knitted

There is increase in the demand of knitted apparels while the concentration of the India's textile trade is more on manufacturing woven fabric. This is also reflected in increasing import of knitted fabric. During the survey, the fabric manufacturers were asked to compare woven and knitted fabric against the various parameters important for deciding for manufacturing woven or knitted fabric. The responses are on a scale of 1-5 (1 being least important and 5 being most important). The weights were assigned and weighted average score are shown in the (Table 5.28 & Exhibit 5.16).

The analysis of sample profile indicates that the majority of respondents are manufacturing woven fabric (Table 5.27).

Table 5.27: Sample profile of fabric manufacturers

Composition of fabric manufactured	Percentage of respondents
100% Woven	84.85
> 50% Woven rest knitted	7.58
100% Knit	7.58

The responses of fabric manufacturers towards reasons for manufacturing woven/knitted fabric are shown in Table 5.28 & Exhibit 5.16.

Woven fabric is perceived to have better (more) market demand, margin, importance of brand image, technological competence and raw material availability. The knitted fabric is perceived to have better availability of finishing facilities, favorable government policies and lesser investment required while on all other parameters discussed the woven fabric is perceived to have better score leading to manufacturers opting for manufacturing of woven fabric rather than knitted one.

Table 5.28: Reasons for manufacturing woven/knitted fabric

Source/Parameters	Woven fabric	Knitted fabric
Market demand	4.47	3.43
Margin	3.64	2.83
Brand image	3.19	2.50
Technology competence	3.62	3.00
Raw material availability	4.03	3.86
Investment required	3.32	3.86
Favorable govt. policies	2.95	3.67
Availability of finishing facility	3.93	4.29

Exhibit 5.16: Reasons for manufacturing woven/knitted fabric

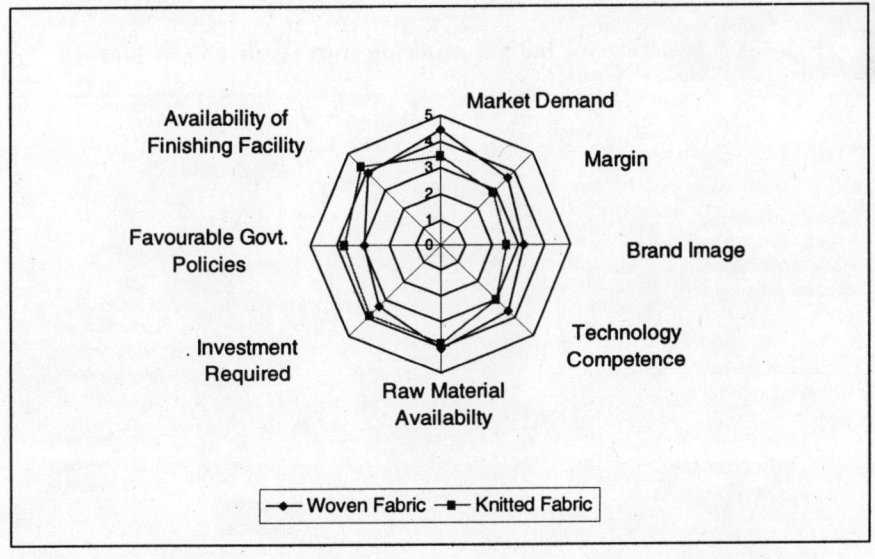

5.3 REASONS FOR NOT ENTERING INTO FABRIC EXPORT MARKET

There are fabric manufacturers only catering to domestic fabric market and not exporting the product. Since the post-MFA period shall provide the opportunity in fabric exports field too. The reasons for not venturing in to fabric exports were ascertained while surveying fabric manufacturers. The sample profile includes 60.87 percent of respondents catering to domestic market only while 39.13 percent of respondents are exporting fabric.

Table 5.29: Reasons for not entering into fabric export market

Reasons for not exporting fabric	*Average Score*
Non compatible quality	2.62
High cost of production	3.52
Higher lead time	3.57
Production of lesser fabric width against demand	2.55
Requisite technology is not available	3.07
Limited production capacity	3.55
Adverse Govt. policy	2.82
Poor processing facility/quality	3.29
Lack of market understanding	3.89
Lack of designing skills to interpret international forecast & demand	3.82
High minimum production quantity	2.88

Exhibit 5.17: Reasons for not entering into fabric export market

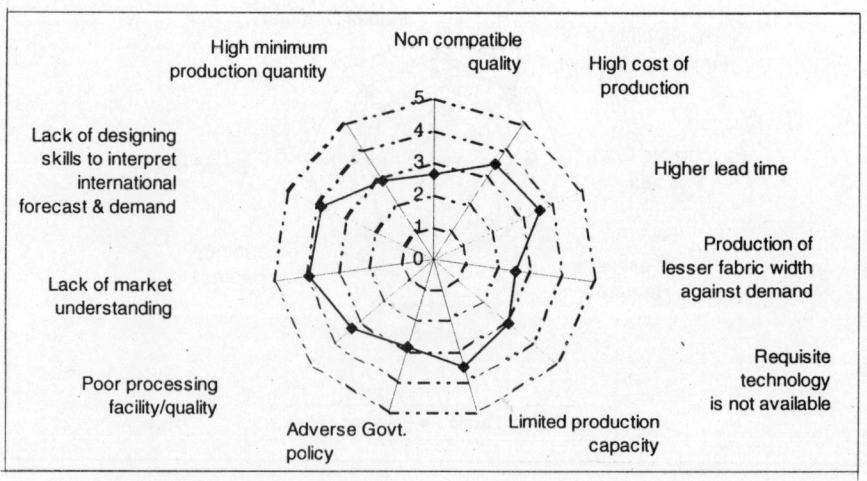

Since in post-MFA period the fabric export is also expected to increase and the manufacturers may increase their business by venturing into exports. The response as shown in Table 5.29 and Exhibit 5.17 indicate lack of market understanding and lack of designing skills to interpret international forecast, limited production capacity followed by higher lead time, high cost of production as the key reasons restricting the entry to fabric export. The competitiveness of Indian textile Industry can be increased by increasing fabric export; the need is for having proper understanding of market alongwith interpretation of fashion forecasts so as to cater to international market in time.

5.4 LESSER FOCUS ON MANUFACTURING HIGH VALUE FABRICS

The opinion of fabric manufacturers was taken to ascertain the reasons for their lack of focus on high value fabrics. The response indicate high cost of manufacturing and limited R&D facility as the key reasons behind manufacturers not manufacturing high value fabric, lack of awareness about new international standards/innovations, higher competition and lack of understanding of market requirement, less profit margin and limited finishing facilities available for high value fabrics are the other reasons behind lesser focus of fabric manufacturers on high value items. The response of fabric manufacturers are shown in Table 5.30 & Exhibit 5.18. High cost of manufacturing for targeting to upper end of the market associated with limited R&D facility and perceived high competition in high quality fabrics are the key reasons leading to lesser focus on manufacturing of high quality fabric in India. The response indicate that availability of raw material for high value fabrics, availability of design skills and skilled workers to produce quality product are not the reasons for lesser focus on manufacturing high value fabric. Further to it, production capacity and availability of requisite technology for manufacturing high value fabric are also not the important factors behind lesser focus on high value fabrics.

Table 5.30: Reasons for lesser focus on manufacturing of high value fabrics

Statements	Average
Reqd. design skills are not available	2.20
Reqd. quality raw material is not available	2.16
Workers are not skilled enough to produce quality product	2.36
Production capacity is not enough	2.44
Requisite technology is not available	2.57
Lack of understanding of market requirement	3.63
Lesser demand in the market	2.92
Less profit margin	3.71
Limited finishing facilities	3.52
Limited R&D facility	4.02
Lack of awareness about new international innovations/standards	3.89
High cost of manufacturing	4.15
Higher competition	3.88

Exhibit 5.18: Reasons for lesser focus on manufacturing of high value fabrics

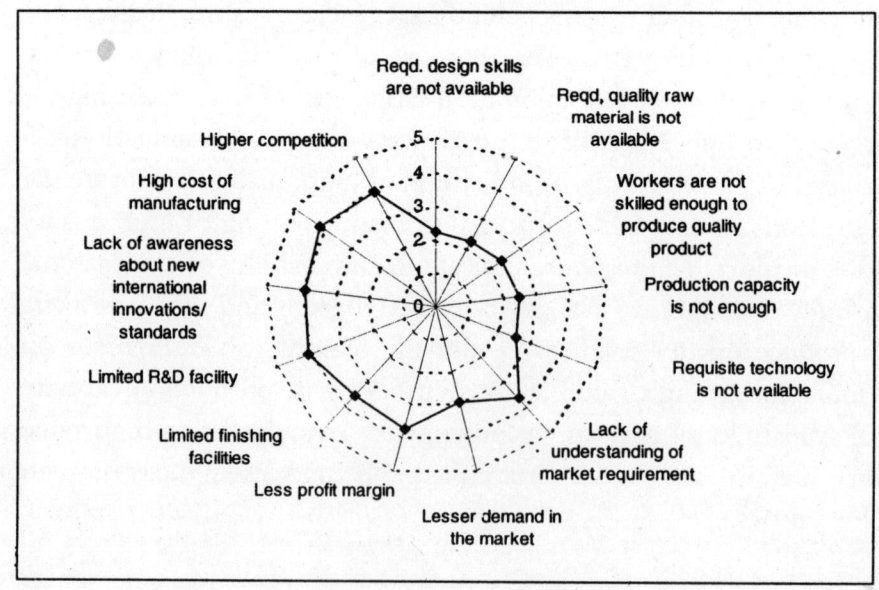

Statistical Analysis

The response of fabric manufacturers is analyzed with factor analysis. Factor analysis using 'Principal Component Analysis' (PCA) was carried out on the data collected, in order to bring out the salient features of the fabric manufacturer's response towards lesser focus on high value items. Table 5.31 indicates factor loadings for various parameters under group 1, 2. The variables alongwith high factor loadings in group 1 include workers are not skilled enough to produce quality product (0.866), required quality raw material is not available (0.834), required design skills are not available (0.773), production capacity is not enough (0.772) and requisite technology is not available (0.694). The variables with high factor loading in-group 2 include limited R&D facility (0.859), lack of awareness about new international innovations/standards (0.845), limited finishing facilities (0.716) and lack of understanding of market requirement (0.682)

Table 5.31: Factor groups for lesser focus on manufacturing high value fabrics

Parameters	Group 1	Group II
Reqd. design skills are not available	.773	.322
Reqd. quality raw material is not available	.834	.245
Workers are not skilled enough to produce quality product	.866	.204
Production capacity is not enough	.772	.335
Requisite technology is not available	.694	.417
Lack of understanding of market requirement	.419	.682
Lesser demand in the market	.438	.564
Less profit margin	.419	.515
Limited finishing facilities	.302	.716
Limited R&D facility	.139	.859
Lack of awareness about new international innovations/standards	.195	.845
High cost of manufacturing	.384	.536
Higher competition	.381	.490

The total variance attributable to the first factor group is 52.99 percent. The total variance for factor group II is 62.60 percent. The output (Table 5.32) indicate that the variables termed under market and R&D functions are highly inter correlated and constitute the key background of lesser focus on manufacturing of high quality fabric. The variables, lack of awareness about new international innovations

(Factor loading 0.859) and high cost of manufacturing (Factor loading 0.845) with high factor loading indicates more importance of these in lesser focus on manufacturing of high quality fabrics.

Table 5.32: Reasons for Lesser focus on manufacturing of high value fabrics

Variables	Factor loading	Factor group title
Lack of awareness about new international innovations/standards	0.859	
High cost of manufacturing	0.845	
Limited R&D facility	0.716	Market and R & D functions
Lack of understanding of market requirement	0.682	

The other variables with high factor loading are Limited R&D facility (Factor loading 0.716) and Lack of understanding of market requirement (Factor loading 0.682). Table 5.33 indicate that the variables; availability of skilled workers to produce quality product (Factor loading 0.866) and availability of required quality raw material (Factor loading 0.834) with high factor loading indicates more importance of these in lesser focus on manufacturing of high quality fabrics. The other variables with high factor loading are availability of required design skills (Factor loading 0.773), enough production capacity (Factor loading 0.772) and availability of required technology (Factor loading 0.694) indicates that Indian industry has a strength in these variables as perceived by fabric manufacturers but due to market and R & D related functions as explained above there is lesser focus on manufacturing of high quality fabric.

Table 5.33: Reasons for lesser focus on manufacturing of high value fabrics

Variables	Factor Loading	Factor Group Title
Availability of skilled workers to produce quality product	0.866	
Availability of reqd. quality raw material	0.834	Production and quality functions
Availability of reqd. design skills	0.773	
Enough production capacity	0.772	
Availability of required technology	0.694	

5.5 SUMMARY

It can be summarized that the fabric from organized sector as well as from unorganized sector in India is perceived to have poor performance in comparison to fabric from imported sources. There is a perception of buying houses that the imported fabric is superior to Indian fabric while making decision of sourcing the fabric for apparel firms. The similar findings are observed while analyzing data (secondary) of import of fabric to India, which indicate increasing import of MMF. This certainly has affected the competitiveness of Indian fabric industry and may further affect it, if, the necessary action are not been taken by fabric manufacturers. The fabric manufacturers are required to improve their perception against the key parameters considered while making a sourcing decision of fabric. The statistical (factor) analysis of responses for various source of fabric for apparel indicates that the criteria for sourcing of fabric are determined primarily by quality functions represented by various parameters (variables). This is followed by price and delivery function. The analysis further indicate that the key variables determining source of raw material for apparel is based upon assessment of quality of various options available to apparel exporters.

The composition of the trade indicates concentration on woven category where as the demand is increasing for knitted categories in last few years. Since the earnings from woven categories are comparatively better, the manufactures are only continuing with work they were doing earlier rather moving in to new direction of trade i.e. manufacturing of knitted categories.Indian fabric manufacturers are targeting to market of greige fabric while there is more market of processed fabric and that is with better price realizations. The reason can be attributed to outdated technology of processing, lesser investment in processing facilities leading to non-availability of required finishing facilities. The fabric export market offers good opportunity but the percentage of manufacture targeting to export market are rather less due to the perception of manufacturers that the market understanding for overseas is non existent and there is lack of design skills to interpret international forecast and demand.

In post-MFA scenario there is wide scope of market expansion but due to increasing competition there is need for moving into high value fabric categories; this shall help in better realization as well as staying competitive in world market. The Indian fabric manufactures have not been able to focus on high quality fabrics due to limited availability of research and development facility and high cost of manufacturing associated with it. The fabric manufacturers also perceive that there is high degree of competition in high quality fabric market in world textile trade. The results of statistical (factor) analysis also indicate that market and R&D function alongwith production &quality function are responsible for lesser focus on manufacturing of high quality fabrics.

Chapter **6**

Post Quota Period: Opportunities for India

The phasing out of the quotas is presenting opportunities as well as posing challenges to the textile and apparel industry. There are many areas with scope for expansion of trade while other areas have emerging challenges in post-MFA period. The strategies for WTO era are to be formulated with the assessment of opportunities as well as challenges. This study shall provide understanding of areas where emphasis is required in post-MFA scenario and where planning need to be done for making Indian textile and apparel industry competitive in world textile and apparel trade. The response of i) Apparel exporters, ii) Fabric manufacturers and iii) Buying houses were taken to understand the above and has been summarized on the following sub-themes.

1. Products expected to be important
2. Impact of increasing fabric import
3. Requirements to increase fabric export
4. Initiatives to be taken by exporters
5. Suggestions to government for policy intervention
6. Summary

6.1 PRODUCTS EXPECTED TO BE IMPORTANT

The response from apparel exporters and buying houses were taken and the percentage of respondents has been indicated along with categories in Table & Exhibit 6.1 – 6.6. The relative extent of expectations for each category by the respondents is depicted by radar diagram. The position of point far form origin in figure has higher response and hence better scope or more opportunity in WTO era.

i) Apparel categories

T-shirts, Gent's shirts, Ladies blouses, Ladies dresses and Trousers are the categories with almost similar response from respondents showing the future potential for them after quota removal. However, there are categories i.e. Ladies skirts, Jackets where there are comparatively lesser expectations from the buying houses for expansion of trade from India in post-MFA period (Table & Exhibit 6.1); apparel exporters expect to have good market for Ladies skirt too. Jackets category is not expected to be much important for India. The reasons can be attributed to non-availability of fabric for Jackets, Outerwear at competitive price. As a matter of fact, specialized fabrics are increasingly being imported. Since

Table 6.1: Apparel categories

Apparel Categories	Apparel Exporters	Buying Houses
Ladies Blouses	13.71	15.75
Gents Shirts	15.73	18.49
Ladies Dresses	16.18	11.64
Ladies Skirts	13.71	7.53
Trousers	13.93	17.81
T- Shirts	15.28	22.60
Jackets	9.89	6.16

*Any other (includes, Babies/Children wear, Sports ware, Sweater, Underwears
Note: All the figures indicate percentage of respondents

Exhibit 6.1: Apparel categories

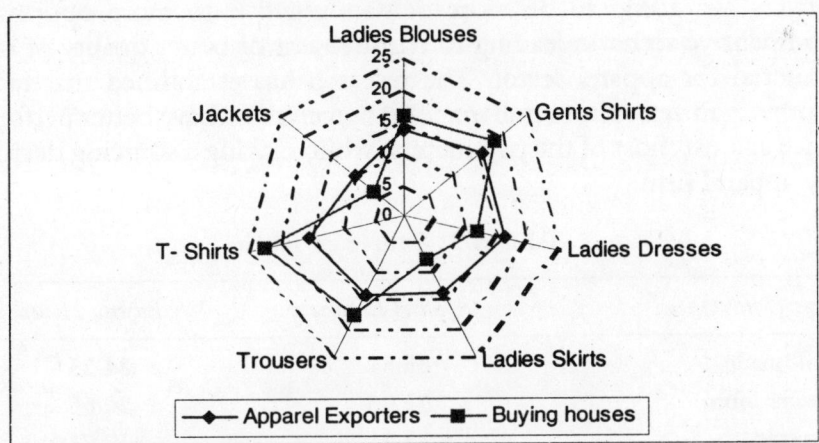

a substantial percentage of cost of apparel is of fabric; Indian apparel industry has an opportunity in categories where it has good raw material base. The textile (fabric) industry needs to expand its base and offer various compositions of fabric as required by apparel exporters at competitive rates.

ii) Value addition in apparel

Apparel exporters have expressed their opinion that the value addition in terms of embroidery, printed and bead work followed by appliqué shall provide growth opportunity in post-MFA period for apparel trade. The buying houses indicated zari, bead work, schiffili and printed as type(s) of value addition to become important in post quota period. The analysis reflect that embroidery, bead work and prints are type of value addition being indicated by apparel exporters as well as buying houses for exports from India in WTO era.

iii) Composition

100 percent cotton followed by cotton polyester blend is the fibre composition with an emerging potential in post-MFA period. The other synthetic fiber blends are also expected to provide opportunities in post- MFA period. This response is associated with India's focus on 100 percent pure cotton and it's expected to continue however as indicted the higher growth in world trade is expected in MMF.

iv) Source of fabric

The analysis of findings of the survey of apparel exporters and buying houses indicate that mill-made fabric is expected to be important in post-quota period followed by knitted fabric. As one is aware that the buyers are going to be more demanding due to more options in competitive scenario leading to requirement of better quality of raw-material for apparel sector. The research has established that fabric from organized sector (mill made) is perceived to have better perform-ance against most of the parameters, while making a sourcing decision by apparel firm.

Table 6.2: Source of fabric

Type (Source)	Apparel Exporters	Buying House
Mill made	35.51	34.25
Power loom	21.96	20.55
Knitted	32.71	39.73
Handloom	7.01	2.74

* All the figures indicate percentage of respondents.

Exhibit 6.2: Source of fabric

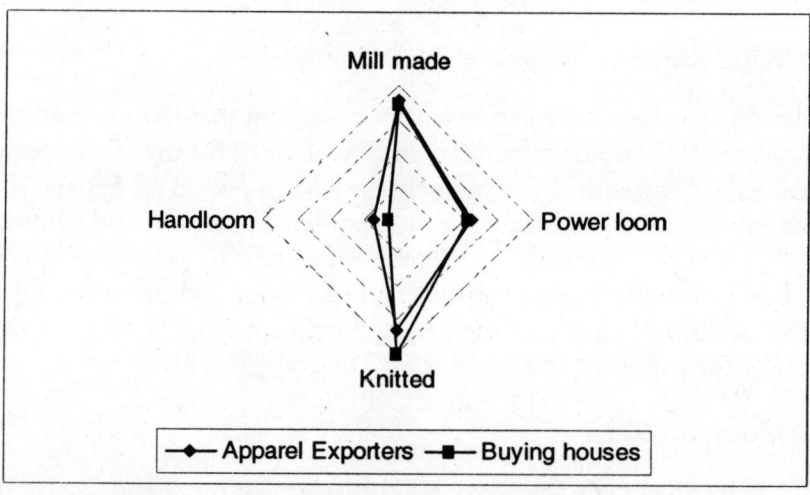

v) Type of fabric

In post-quota period processed fabric (printed, dyed) are expected to be important. The potential for greige fabric is expected to be lesser in comparison to finished one. The findings in Table 6.3 indicate oppor-tunities hence provide direction to Indian fabric manufacturers for post-MFA period.

The world trade is moving towards processed fabric but India's trade is primarily of greige fabric which leads to lower price realizations and less competitiveness of Indian textile industry due to its focus not being on value added (i.e. processed) fabric. The Indian textile industry is required to concentrate on trade of processed fabric to compete in post quota world trade and have better price realizations. Investments are required in processing facilities by Indian textile industry.

Table 6.3: Type of fabrics

Type of fabric	Percentage of respondents
Greige	10.39
Bleached	10.39
Dyed	25.97
Yarn Dyed	25.97
Printed	20.78

*All the figures indicate percentage of respondents

vi) Fabric width

In post-MFA period, greige fabric width (61-80") followed by (40 – 60") and finished width (40-60") followed by (61-80") is expected to provide more opportunities to Indian fabric manufacturers (Table 6.4) Since wider width fabric provide economy in production of apparel items and hence preferred by apparel manufacturers. This is important to note that production of wider width fabric is possible only with latest technology looms. India lacks in modern technology looms leading to poor performance of weaving sector in India. Since the more opportunities are indicated in trade of wider width of fabric, efforts are required to improve technological performance of weaving sector. The fabric being imported from China, Korea etc. of wider width and is preferred over lesser width fabric for items of mass production where economy in fabric consumption highly affects costing.

Table 6.4: Fabric widths

Widths	Greige	Finished
Can't say	27.27	10.34
40-60"	27.27	44.83
61-80"	31.82	27.59
80-100"	4.55	6.90
100-120"	4.55	6.90
> 120"	4.55	3.45

6.2 IMPACT OF INCREASING FABRIC IMPORT

Since the textile and apparel sector is getting liberalized with the phasing out of the quotas the import of fabric is also expected to increase. The increasing fabric imports to India may provide a challenge to domestic fabric manufactures. The response of fabric manufacturers was taken to understand their viewpoint on the issue of the impact of increasing fabric import.

A majority of respondents have expressed (Table 6.5) that either the impact is somewhat significant to significant for the domestic textile industry. The impact of increasing fabric import is expected to be more

Table 6.5: Impact of increasing fabric import

Response	Percentage of respondents
Not at all	1.67
Insignificant	10.00
Somewhat Significant	45.00
Quite significant	28.33
Very damaging	10.00

on power loom sector (56.98 percent of respondents) followed by mill sector (31.40 percent of respondents), as the quality norms are expected to be more stringent. Blended as well as 100 percent man-made fabrics are expected to get affected due to increasing imports. The effect is less likely to be felt by cotton material due to India's strong base in cotton. The findings also indicate that the impact will be more visible on finished fabric rather than greige fabric. The manufacturers of greige fabric width 61-80" and finished fabric width of 40-60" are expected to be affected more in post-MFA period.

The fabric manufacturers were asked to indicate the parameters, which may lead them to effectively compete with imported fabric. The responses were taken for various parameters on a scale of 1-5 (where 5 is extremely important & 1 is not at all important). The weights were given and average score are shown with radar diagram. The responses indicate that quality of final product, raw material quality followed by cost control are the key parameters for Indian industry to effectively compete with imported fabric in post-quota period followed by the other factors i.e. lead time, exclusivity in design and investment in latest technology alongwith scale of production (Table 6.6 & Exhibit 6.3). It is felt that final quality of product is an important criterion to face competition with imported fabric. The requirement of good quality raw material with cost competitiveness highlights the importance of strengthening the elements of value chain of textile industry. It is also important to have latest and exclusive design at right time in market. There is a need for investment in R&D as well as latest technology for better quality product.

Table 6.6: **Requirements to compete with imported fabric**

Parameters	Average
Scale of production	3.86
Investment in latest technology	4.04
Quality of product	4.41
Raw material quality	4.35
Cost control	4.33
Any other (lead time, exclusive design, taxation policy etc.)	4.40

Exhibit 6.3: **Requirements to compete with imported fabric**

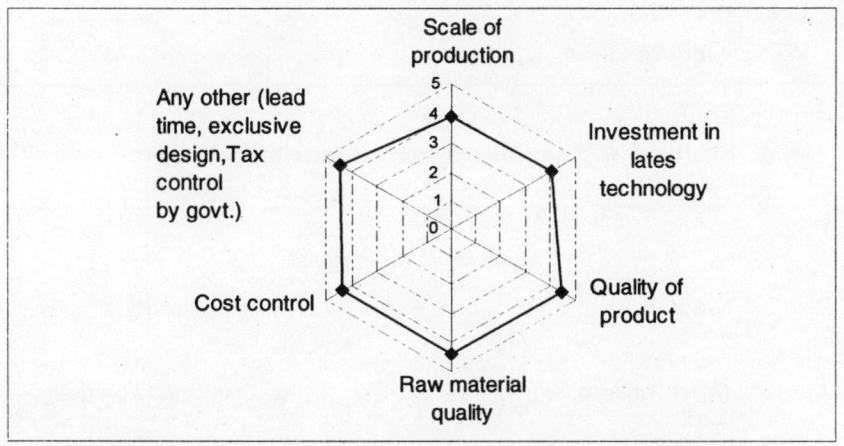

6.3 REQUIREMENTS TO INCREASE FABRIC EXPORT

The post-MFA period shall provide opportunity for expansion of textile and apparel trade. There is potential of increasing fabric exports to the destination markets. The responses were taken from fabric manufacturers on the parameters felt to be important for increasing fabric exports.

The parameters are in the scale of 1-5 (1 being least important & 5 being most important). The weights were assigned and weighted average scores are indicated in table and shown with the radar diagram. The sample profiles for fabric manufacturer's survey indicate that only 39.13 percent of the respondents are exporting fabric while 60.87 percent of manufacturers are not exporting fabric.

The analysis (Table 6.7 & Exhibit 6.4) indicates quality of end product as the most important parameter for increasing exports followed

Table 6.7: Requirements to increase fabric export

Important parameters for increasing exports	Average
Availability of raw material	4.22
Production capacity	3.81
Cost competitiveness	4.52
Management attitude	4.07
Export market information	4.23
Duty structure	3.64
Quality of end product	4.63
Technological competence	4.43
Order (lot) size	3.23
Wider width of fabric	2.72

Exhibit 6.4: Requirements to increase fabric export

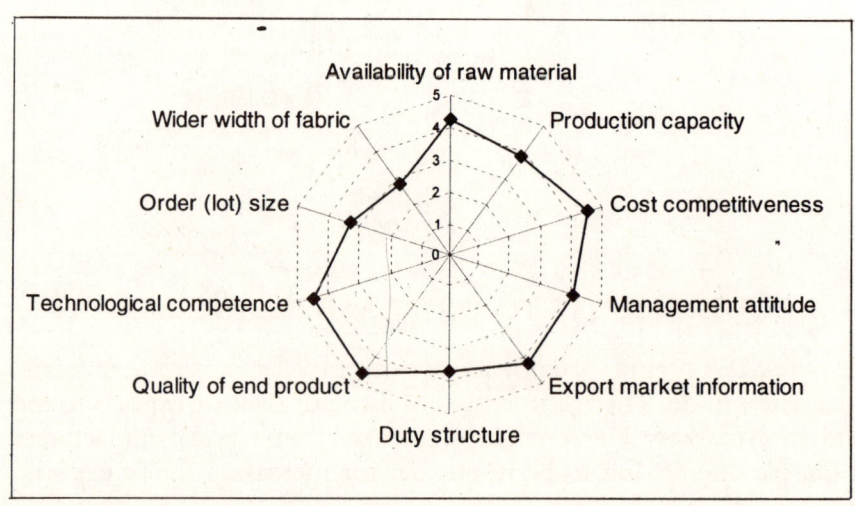

by cost competitiveness, technological competence. The knowledge of export market, availability of raw material and management attitude are also important parameter contributing in increasing exports. A further closer look indicates that quality of end product is linked with availability of raw material and technological competence. This coupled with cost-competitiveness play an important role in determining competitiveness of a textile firm in World trade.

6.4 INITIATIVES TO BE TAKEN BY EXPORTERS

As discussed earlier, the post-quota period is understood to have increased competition with demanding customers due to more options with them, it is important for industry to take steps to face competition in international market. The areas with increasing opportunities are analyzed in section 6.1. The apparel industry is required to pursue strategies to be competitive in world trade in emerging areas. The response for the above has been taken from apparel exporters and buying houses separately. The response against various steps to be taken by the exporters, has been collected on a scale of 1-5 (1 being least important; 5 being most important). The weights were assigned and weighted average score are shown in Table (6.8 – 6.10) and Exhibit (6.5-6.7).

The response of apparel exporters towards various parameter as initiative to be taken in post quota period is shown (Table 6.8 & Exhibit 6.5). The response of buying houses (Table 6.9 & Exhibit 6.6) indicate importance of various initiative to be taken by exporters. The comparison of responses from apparel exporters and buying houses (Table 6.10 & Exhibit 6.7) summarises relative importance given to various initiatives by exporters and buying houses.

6.4.1 Response of Apparel Exporters

The analysis of the response of apparel exporters indicate (Table 6.8 & Exhibit 6.5) the requirement of focus on product development and design followed by the need for investment into latest technology as foremost important measures to face competition in International market. Besides it, focus on better source of raw material, product specialization and extensive marketing of the products, increase in production capacity are the other important parameters requiring attention and action from apparel exporters. There is lesser importance of requirement of launch of own brands by exporters.

Table 6.8 Initiatives to face competition in international market

Parameters	Average
Launch own Brands	2.87
Extensive marketing of the product	4.47
Need for consortium of manufacturers	4.28
Product specialization	4.46
Investment into latest technology	4.50
Diversification into new product categories	4.23
Increasing of product capacity	4.42
Add backward/forward linkage	4.29
Focus on better source of raw material	4.48
Focus on product development and design	4.54

Exhibit 6.5: Initiatives to face competition in international market

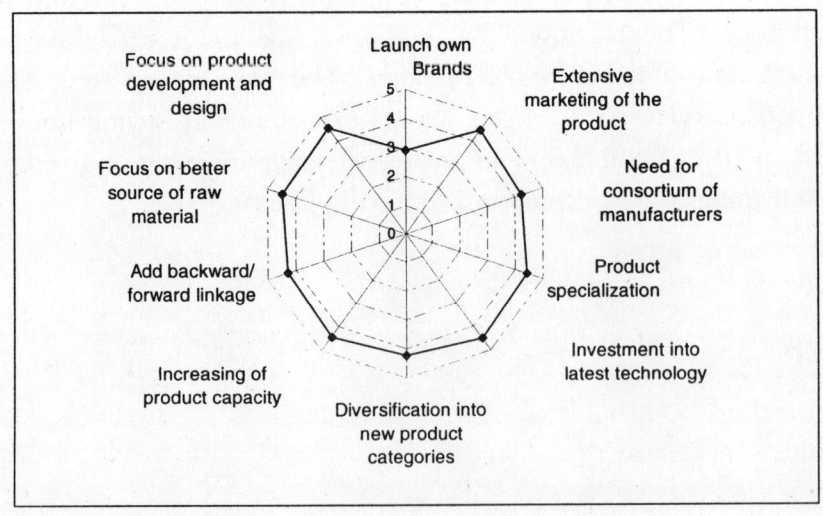

6.4.2 Response of Buying Houses

The response of buying house indicate focus on product development and design as the foremost important parameter followed by investment in the latest technology, focus on better source of raw material as the important step to be taken to be competitive in world trade. Product specialization, increase in production capacity are other steps to be taken by Indian manufacturers and exporters in this respect. The response indicate lesser importance of launch of own brand of manufacturers need for consortium of manufacturers and adding backward linkages to the capacity (Table 6.9 & Exhibit 6.6).

Table 6.9: Initiatives to face competition in international market

Parameters	*Average*
Extensive marketing of the product	3.84
Launch of own brand	2.54
Need for consortium of manufacturers	2.70
Product Specialization	4.41
Investment into latest technology	4.59
Diversification into new product categories	3.80
Increasing production capacity	4.13
Add backward	3.40
Focus on the better source of raw material	4.50
Focus on product dev. And design skills	4.69

Exhibit 6.6: Initiatives to face competition in international market

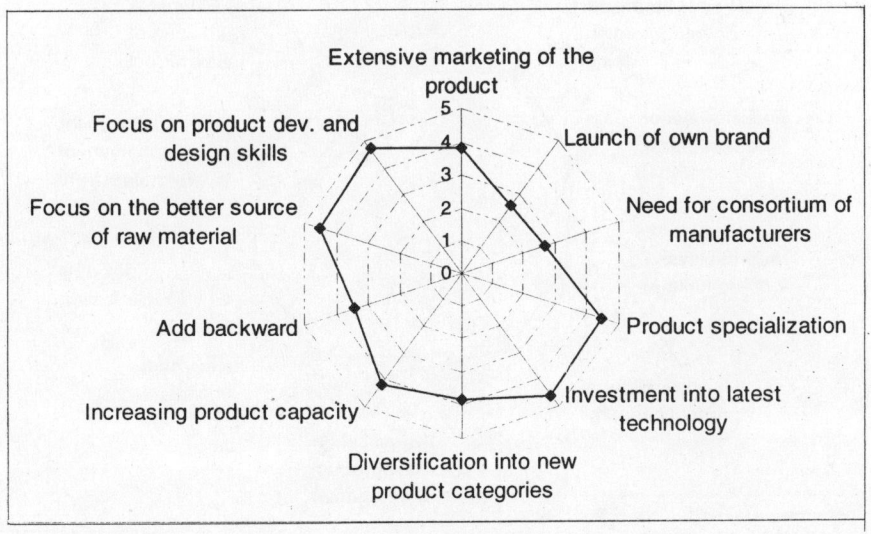

6.4.3 Comparison of responses from Apparel Exporters and Buying Houses

A comparison of responses from apparel exporters and buying houses is shown in (Table 6.10 & Exhibit 6.7) which indicate focus on product development and design skills, investment in to latest technology as important step followed by focus on better source of raw material,

Table 6.10: Perception gap between apparel exporters and buying houses

Parameters	Apparel exporters	Buying houses
Launch own Brands	2.87	3.84
Extensive marketing of the product	4.47	2.54
Need for consortium of manufacturers	4.28	2.70
Product specialization	4.46	4.41
Investment into latest technology	4.50	4.59
Diversification into new product categories	4.23	3.80
Increasing of product capacity	4.42	4.13
Add backward/forward linkage	4.29	3.40
Focus on better source of raw material	4.48	4.50
Focus on product development and design	4.54	4.69

Exhibit 6.7: Perception gap between apparel exporters and buying houses

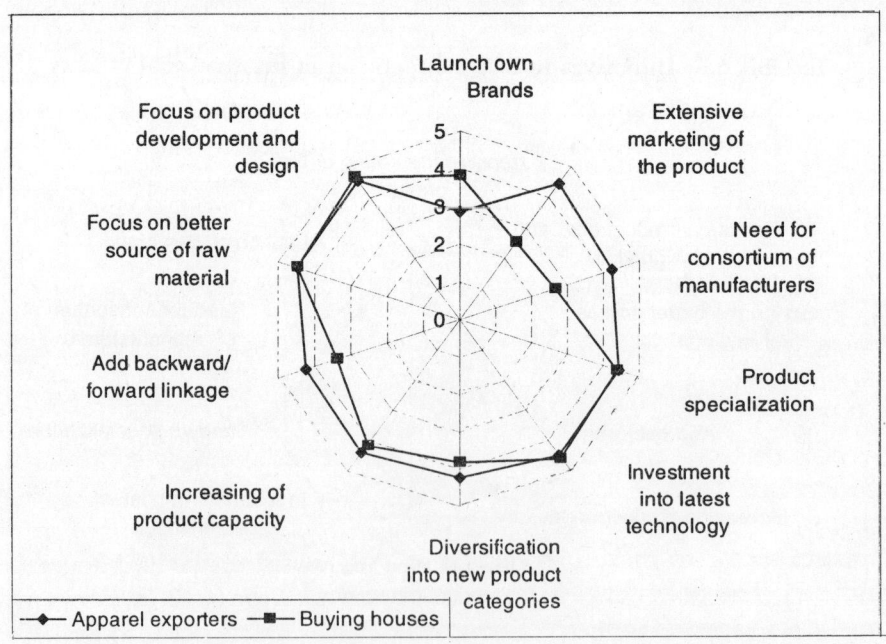

product specialization as important step to be taken by apparel firms to be competitive in world market. Diversifications in to new product categories, addition of backward/forward linkage is also important initiative in this direction to be taken by apparel exporters. There is lesser requirement of launching own brand, need for consortium of manufacturers while exporters felt requirement of extensive marketing along with need for having a consortium of manufacturers.

6.5 SUGGESTIONS TO GOVERNMENT FOR POLICY INTERVENTION

The suggestions were taken from apparel exporters, fabric manufacturers and buying houses so as to understand expectations of various stake-holders in the textile and apparel trade from government for facing competition in WTO era.

The prevalent duty structure for textile and apparel trade need to be rationalized in view of post-quota scenario and the duty structure need to be at par or competitive with the competing countries i.e. China, Bangladesh etc. The labour laws of India are considered to be very stringent and there is a demand from the exporters to have flexible labour laws. The section of the exports felt that the labour laws for the units engaged in export should be different then other industry. Some of the manufacturers and exporters have asked for allowing retrenchment of labour due to cyclic nature of the fashion industry. There is a wide & strong feeling against prevalent bureaucratic system having extensive paper work, which creates hurdles. The efforts are required to make paper work simpler and the government authorities need to be more responsive to the manufacturers and help them by reducing red tapism. The manufacturers and exporters have also asked for more subsidy to be competitive with the neighboring countries i.e. China. Infrastructure in terms of road, water and transportation (road, air and rail) need to be improvised so as to enable manufacturers and exporters in cutting down the lead times and cost which play a crucial and important role in competition.

The help of government is also required for seeking more information of target markets for new market development as well as to understand the growing business areas. Since smaller manufacturers may not have access to this information, government's help is expected. The policy interventions are required to provide access of imported technology at lesser duty to improve level of technology in textile and apparel sector to make it competitive with world textile and apparel industry. The policy should be attractive enough for large investors in textile and apparel industry. The income from exports needs to be 100 percent tax free so as to lure more manufacturers to venture in exports. The government intervention is also required for making easy availability of fabrics, trims besides latest fashion forecast. The government policy need to be prevalent for a specific period, in other words the policy needs to be stable as the frequent changes pose a threat for stability of industry. The government policies should encourage new investments by providing tax holidays in priority areas i.e. processing.

6.6 SUMMARY

It can be summarized that in world trade T-shirts, Gents shirts, Ladies blouses, Ladies dresses & Trousers are the apparel categories expected to provide opportunities in post-MFA scenario. As discussed earlier there is need for value addition in terms of Zari, bead work and printed apparel. 100 percent cotton followed by blend (Cotton - Polyester) and other blends are fiber composition to be important for India after quota removal. In post-MFA scenario, mill made (woven) fabric followed by knitted fabric are to be more important in Indian textile and apparel Industry. It also indicate processed fabric offering more scope for future alongwith wider width of fabric in WTO era due to increasing competition in world textile and apparel market.

The impact of increasing fabric import in post-MFA period is expected to affect more to powerloom fabric manufacturers but overall textile industry is also expected to be affected .The customers are poised to look for sourcing better quality of raw material. The MMF sector shall be widely affected due to increase in import owing to less specialization of Indian textile industry and less cost competitiveness in this area. Wider width fabrics are to be preferred with increasing fabric import by apparel manufacturers due to economy in production and ease of operations. It will affect powerloom as well as mill sector (fabric) due to old technology with Indian textile industry. The imported fabric shall primarily affect greige fabric manufacturers. Indian fabric manufacturers are required to improve quality of product by improving raw material quality and investment in latest technology while having cost control to remain cost competitive and face competition from imported fabric by offering latest and contemporary design as per customers' requirement.

Besides it, fabric exports shall also provide opportunity to manufacturers. To be competitive as fabric exporter, quality of end product needs to be prime concern with high degree of cost competitiveness. The Indian industry needs to invest so as to achieve technological competence & have investment in value chain to have better raw material. Indian apparel exporters are required to focus on product development and design, better source of raw material besides product specializations. The investment into latest technology is also required to be competitive in world. While the textile and apparel industry need to take various steps to be competitive in world, the government also needs to intervene by having a competitive policy frame work in tune with other competing countries in the world.

Chapter 7

Strategies for Success in WTO Era

The textile and apparel industry is one of the largest industrial sectors in India and a leading foreign exchange earner. Industrial liberalization in the domestic economy since 1991 has been followed by changes in the global environment. Abolition of licensing controls on the industry was followed by the Uruguay round of negotiations resulting in a ten-year phase out of textile export quotas under the 'Agreement on Textile and Clothing'. The roll down of textile quotas was accompanied by roll down of custom duties and quantitative restrictions on import of wide range of textile and apparel products in to the Indian market. These sequential steps are leading to a convergence of the domestic and international markets for textiles and apparel. The domestic industry is no longer insulated from global competition. The phased removal of textile quotas since 1995 has catapulted Indian export firms in to a new competitive environment.

The exports of Indian textile and apparel have grown under the environment of MFA quotas for over two decades. The international textile and apparel trade was conducted in a unique regulatory environment that restricted exports from specific countries and of specific products. There has been considerable speculation and divergence of opinion among industry and government circles on how Indian exports may be affected with removal of MFA quotas. One school of thought is that Indian industry will be freed from quota shackles and able to boost exports. The other school of thought is that MFA quotas provide secured access to US and EU markets which may be adversely affected with removal of quotas. This school of thought is based on the premise that Indian industry may not be able to withstand international competition in a quota free regime. An attempt has been made in this book to assess the sources of competitive advantage for Indian apparel & textile industry in WTO era. The research findings of this study provides an insight on how the removal of MFA quota is likely to influence Indian textile and apparel industry and preparedness of industry for WTO era.

7.1 STRATEGIES FOR TEXTILE INDUSTRY

The research findings indicate high cost of manufacturing, limited R&D facility, higher competition and lack of understanding of market requirement, lack of awareness about new international innovations/ standards as the key reasons for Indian fabric industry not being able to focus on manufacturing high value fabrics in India. The research findings also revealed lesser marketing effort requirement and more demand as the reasons for exporting more of greige fabric than value added processed fabric. This coupled with expertise in the product makes fabric manufacturers continuing with manufacturing of greige fabric despite of increasing demand of finished fabric in world trade. The fabric manufacturers are not being able to have in roads into fabric export due to lack of understanding of target market, lack of designing skills to interpret international forecast, higher lead time and limited production capacity.

The Indian fabric manufacturers are concentrating more on woven categories despite the increasing requirement of knitted fabric in world apparel market due to perception of higher market demand, better margin, easy raw material availability, technological competence & availability of technology for manufacturing woven fabric. The government policy framework reserving knitted under SSI for a larger period also hindered bigger investments in knitted sector.

100 percent cotton followed by polyester cotton blend is expected to be important for India after quota removal. This is also found that processed fabric shall be much more important due to competition with imported fabric. The fabric width of 40 – 60" (finished) and 61-80" (Greige) is to be more important and be in demand after quota removal.

With phasing out of quotas the increase of imports of textile and apparel is expected. It is felt that powerloom sector will be more affected due to stricter quality norms of fabric buyers for apparel industry and easy availability of fabric from imported origin. The impact of increasing import shall be felt more upon synthetic fabric from Asian countries. The impact shall be felt more on processed fabric as imported fabrics are primarily processed. To face the competition from imported fabric, it is required to improve the quality of the product, raw material quality and exercise cost control.

The measures to be taken to increase fabric exports include improving the quality of end product, cost competitiveness, technological competence etc. Cost competitiveness is affected by set of factors consisting raw material, technology & productivity. Due to lesser availability of raw material for synthetic fabric in India and poor state of technology particularly in weaving and processing; India is not cost competitive and the quality of end product of large number of unorganized fabric manufacturers is not suitable for exports.

Exhibit 7.1 shows ideal positioning of various parameters (scale of 1 to 5, 5 being highest) as measures to increase fabric exports. Compatible quality is one of the critical requirements besides low cost of production, lower lead time and production of wider width fabric. To

Exhibit 7.1: Measures to increase fabric exports

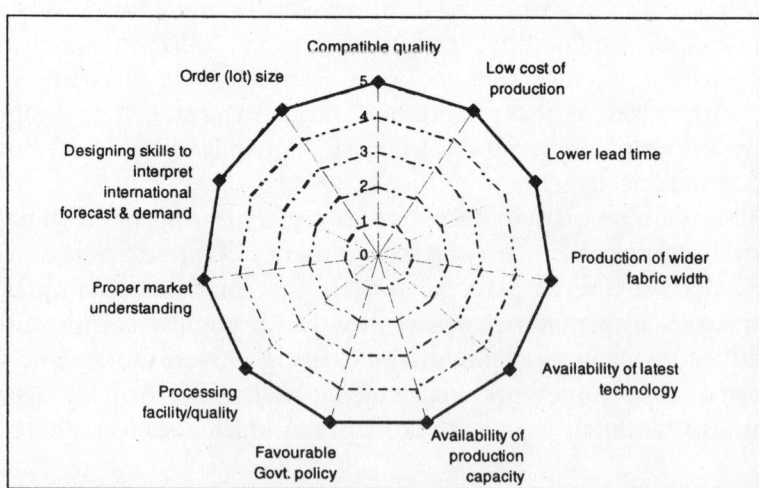

achieve it, the need is for availability of latest technology, processing facilities with quality output besides availability of production capacity and flexibility of catering to different and larger order (lot) size. The industry needs proper and timely market understanding for destination and skills to interpret international forecast and demand and cater it. The government policy is required to be favourable to industry so as to enable it to increase fabric exports in WTO era. The new policy initiative are to be comparable with competing countries in world textile trade.

The fabric from organized sector as well as from unorganized sector in India is perceived to have poor performance against various parameters in comparison to fabric from imported sources leading to perception of export houses and buying houses that the imported fabric is superior to the Indian fabric while making decision of sourcing the fabric for apparel firms. This is causing increasing import of fabric (particularly MMF) to India. This certainly has affected the competitiveness of Indian fabric Industry and may further affect it, if the necessary action is not been taken by fabric manufacturers.

The increase in consumption of imported fabric is the result of non-availability of desired fabric in domestic market. India is perceived to be preferred source of 100 percent cotton based fabric while the fabrics with MMF are being increasingly imported from China, Taiwan, Korea. The increasing import of fabric is due to better performance of the fabric in terms of availability of wider width, consistency in quality, required quality of finishing, consistency in lot/roll quantity, lot/roll quantity availability and count and construction availability for required fabric.

Exhibit 7.2 shows ideal position (scale 1 to 5, 5 being highest), of various parameters (key consideration for sourcing decision of fabric) to be competitive in WTO era and effectively compete with fabric from imported sources. Indian textiles industry is required to be prepared to deliver high order quantity with consistency in quality and quantity at competitive (lower) price at minimum lead time. The level of technology is required to be equipped to provide wider width of fabric, required quality of processing (including finishing). The lot/roll quantity availability, availability of various counts and construction and required physical property for end uses and low samplings cost are other main requirements for Indian fabric industry to be competitive in WTO era.

Exhibit 7.2: Requirements for competitiveness in WTO era

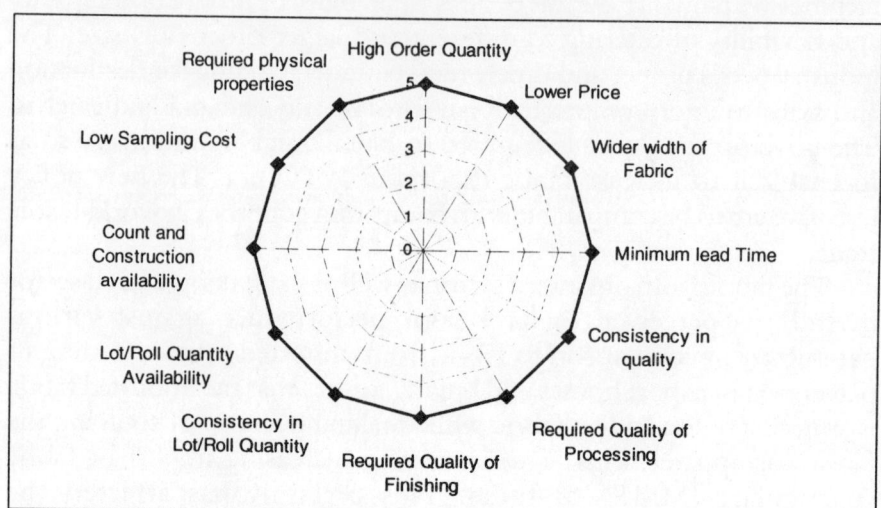

The Indian fabric manufactures have not been able to focus on high quality fabrics due to limited availability of research and development facility and high cost of manufacturing associated with it. The fabric manufacturers also perceive that there is high degree of competition in high quality fabric market in world textile trade. Indian fabric manufacturers are targeting to market of greige fabric while there is a large market of processed fabric available with better price realizations. The reason can be attributed to outdated technology of processing, lesser investment in processing facilities leading to non-availability of required finishing facilities in India.

Exhibit 7.3 shows ideal positoning (scale 1 to 5, 5 being highest) of factors which may reposition India as source of high value fabrics in world market. Indian fabric (textiles) industry is required to have under-standing of market requirements and awareness about new international innovations and standards to cater high value segment in fabric market. The requirements include availability of required design skills, raw material, skilled workers to produce quality product supported by latest technology including finishing, R&D facility. It is also necessary to be cost competitive in market. The high and increasing demand in this segment of market is accompanied by stiff competition, demanding buyers alongwith higher profit margin.

Exhibit 7.3: Requirements for repositioning of India as source of high value fabric

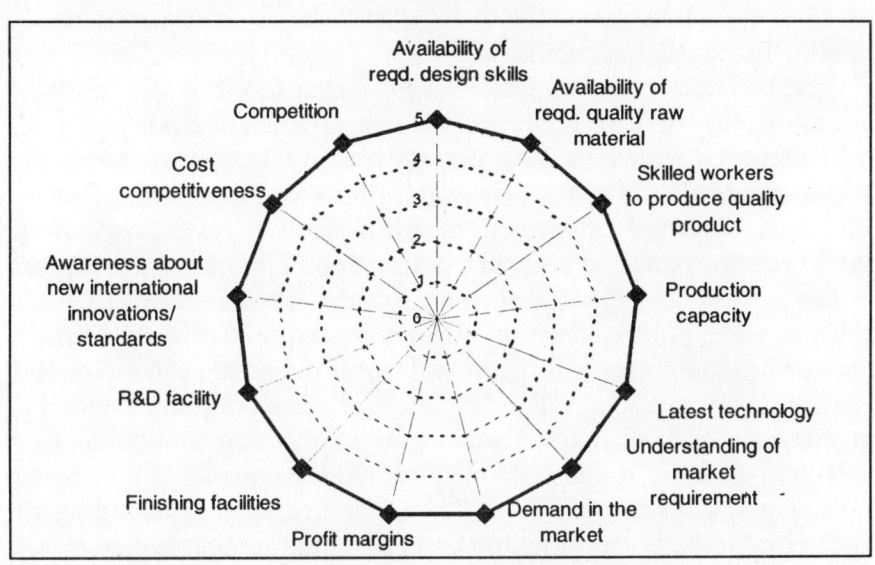

7.2 STRATEGIES FOR APPAREL INDUSTRY

The research findings show that India is targeting to lower end of the market because of lack of differentiation in the value chain configuration for basic and high value items. The apparel exporters are primarily exporting the basic categories i.e. Gent's shirts, Ladies blouses, Ladies skirts and T-shirts etc. The limited research and development facility, non-availability of quotas, unfavorable cost competitiveness alongwith the image of India as destination for sourcing basic items has contributed in it. Technological constraints, limited finishing facilities and difficulty in importing raw material are the other key reasons leading to lesser focus on manufacturing of high value added apparel products in India. The productivity of Indian firms is rather poor in comparison to world, which causes and contributes in unfavorable cost competitiveness of apparel industry. One of the key reasons behind more focus on basic items is non-availability of required fabric for high value items. The lower productivity is also caused due to the poor state of technology, small-scale nature of industry, lesser investment in technology in assembly line production.

There is increasing demand of knitted apparels in world market while the Indian manufacturers are still concentrating more on manufacturing of woven apparels. The analysis of perception of apparel exporters indicates high market demand, margin and quota availability as the reason for interest in woven apparels while technological competence in manufacturing woven categories is also a contributing factor. However, It is perceived that the availability of raw material is superior for knitted apparels.

The increase in consumption of imported fabric is the result of non-availability of desired fabric in domestic market. India is perceived to be preferred source of 100 percent cotton based fabric while the fabrics with MMF are being increasingly imported from China, Taiwan, and Korea. However for synthetic blends India as well as China, Taiwan and Korea are preferred sourcing destination. The increasing import for fabric is due to better performance of the fabric in terms of availability of wider width, consistency in quality, required quality of finishing, consistency in lot/roll quantity, lot/roll quantity availability and count and construction availability. The fabric from organized sector is preferred by apparel manufacturers due to meeting requirement of high order quantity, minimum lead-time, required quality of processing and required physical properties. The fabric from un-organized sector is perceived to have low sampling cost. This indicate that Indian textile (fabric Sector) is not being able to meet the requirement of apparel manufacturers and leading them to focus on identifying better source of raw materials to face competition in international market.

The study indicated the need for focus on product development and design, investment into latest technology, extensive marketing of the product and product specialization alongwith increase in production capacity as measure to face competition in international apparel market.

Ladies dresses, Gents' shirts, T-shirts, Ladies blouses and Skirts are apparel categories having potential for apparel exports in post-MFA period. It is also expected that the value addition i.e. embroidery, print and bead work will be important in WTO era. Since India is having a strong base in cotton, 100 percent cotton is the fabric composition expected to remain important for India after quota phase-out. The findings of survey indicate that limited R&D facility, limited finishing facility and unfavorable cost competitiveness and difficulty in importing raw material are key reasons leading to lesser sourcing of high value items from India. Besides it, the image of India as the producer of basic product and technological constraints are the other reasons for more concentration on sourcing of basic items from India. These parameters include the image of India as producer of basic items, difficulty in importing raw material, unfavorable cost competitiveness and limited R&D facility in India. The image of India as producer or source of basic merchandise is due to concentration of trade in basic categories with average FOB lesser than US$ 4.

Exhibit 7.4 shows pictorial description of requirement for repositioning of India as source of high value itmes on a scale of 1 to 5 (5 being most important). Availability of required design skills, raw material, workers with required skills is necessary for repositioning of India as source of high value items. There is need for latest R&D facility, finishing facility besides ease in importing raw material. Technological constraints have to be minimum and cost competitiveness is to be achieved with combination of all factors re-position India as source of high value items.

Exhibit 7.4: Requirements for repositioning of India as source of high value apparel

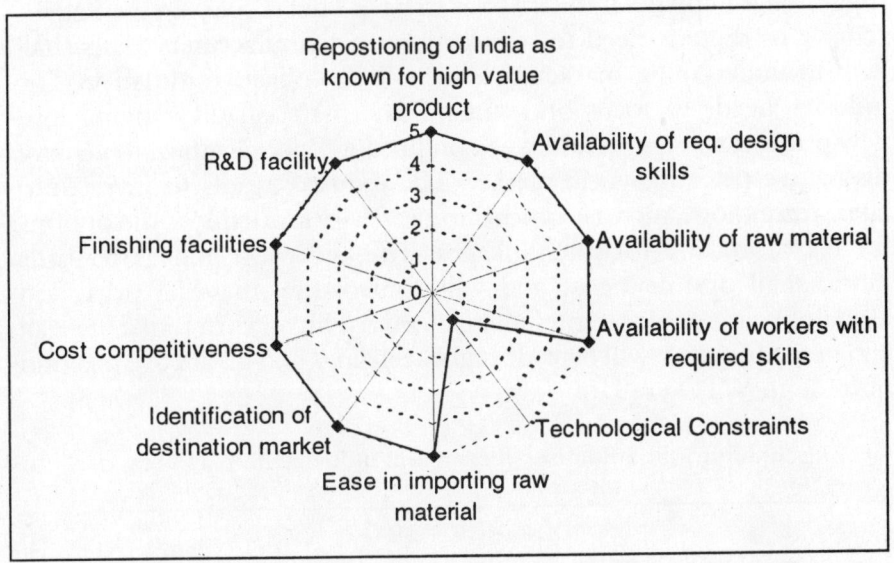

The findings also indicate that although there is good market demand for knitted apparels but more of woven apparels are being sourced from India due to manufacturer's technological competence in woven, availability of quota, availability of production facility and finishing facility. In nutshell, some parameters are more favorable to source woven apparels from India despite of good market demand of knitted apparels with phasing out of quotas and more investment in knitted sector in India; the scenerio is likely to change in coming years.

Focus on product development and design, investment into latest technology and focus on better source of raw material and product specialization are the initiatives to be taken by the Indian apparel industry so as to enable it to face competition in international market. The categories with potential of business expansion in post-2004 include

T-shirt, Gent's shirts, Trousers and Ladies blouses. The value addition is also expected to be important for exports from India to differentiate its offerings & provide better price realization. 100 percent cotton is the key composition to remain important even after phasing out of the quotas. Knitted sector is expected to be important for India in changed scenario in quota free world as the world trade in textile and apparel is shifting towards knitted apparels with changing lifestyle of target customers.

Exhibit 7.5 shows that to be competitive in WTO era, the Indian apparel industry needs to focus on product development and design so as to cater to requirement of target customers. Launching our brands in destination market besides extensive marketing of the product is suggested for apparel industry so as to have a niche space in market. To achieve it, there is need for consortium of manufacturers. This shall help in undertaking marketing efforts with collective initiative. The industry needs to focus on sourcing raw material, diversifying into new product categories and work on product specialization to achieve distinctive place in world market. The industry needs to invest into latest technology, increase production capacity and explore the options for backward and forward linkages. The backward linkage towards fabric shall optimise cost and will control raw material price and availability as per requirement while forward linkages in terms of retail and brand initiative will provide higher earnings and distinctive position in trade in WTO era.

Exhibit 7.5: Initiatives for apparel industry in WTO era

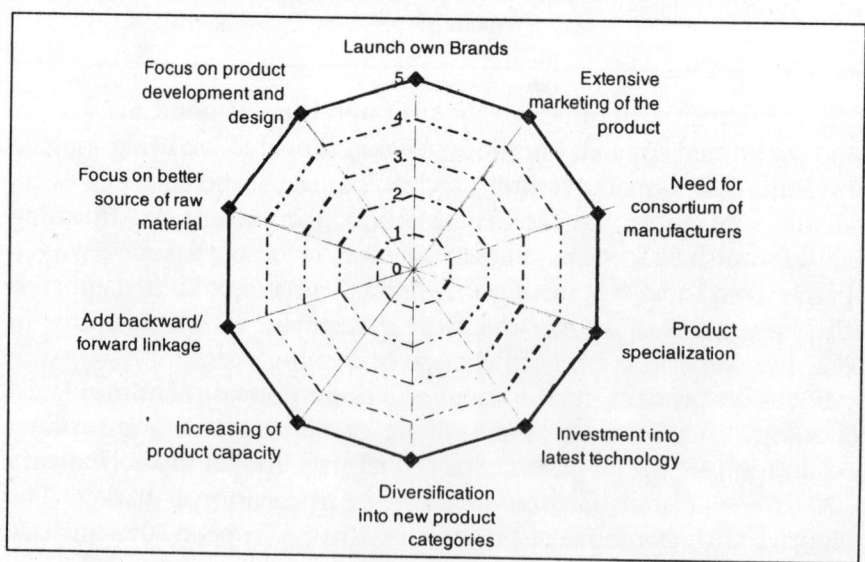

7.3 SUMMING UP

Indian apparel and textile exports has high potential to upgrade its inherent comparative advantages and move towards sustainable competitive advantage especially to prepare for the competitive scenario in the post-MFA textile and apparel trade.

The need is to position India as a supplier of high value items. The apparel exporters are required to make efforts for changing the perception of being producer of basic items. It can be achieved by technological innovation and usage of latest technology in manufacturing and processing. The apparel manufacturers have to use quality raw material (fabric) with value addition to differentiate its offerings and target to upper end of market. To face competition in WTO era the focus on product development & design is needed which requires investment into latest technology and focus on better sources of raw material. Indian apparel manufacturer need to have product specialization to get higher earning. The increasing competition in basic categories shall bring prices down in post-quota scenario. The manufacturers need to invest into latest plants & develop expertise in higher value items.

There is increasing demand of knitted apparels in world market while Indian manufacturers are primarily concentrating on woven apparels. The reason lies in better margin, quota availability & technology competence for manufacturing woven fabrics. The need is to have specialization in knits alongwith woven fabrics and target on higher end of market.

As world textile and apparel trade is shifting to apparel. The textile and apparel industry of a country can remain competitive in quota free world with competitive advantages in each element of value chain of manufacturing. The apparel industry is increasingly becoming dependent upon fabric of imported origin due to perception of better performance on various parameters including lower cost, availability of wider width fabric, consistency in quality, processing & moreover minimum lead time. The fabric from organized sector is perceived to be preferred by manufacturers for higher quantity requirement, lower sampling cost. The fabric form unorganized sector which contributes around 95 percent of India's production, is only preferred for low sampling cost. It indicates that India's textile (fabric) sector needs to become competitive by investment in latest technology in weaving and processing, research and development and offering product with proper understanding of market requirement.

Indian fabric is preferred for 100 percent cotton while for synthetics China, Taiwan & Korea are becoming sourcing base for Indian apparel manufacturers due to cheaper price and superior quality. Indian textile (fabric) industry needs to offer merchandise as per latest international trends. The requirement of apparel industry is for consistency in quality and quantity, wider width and cost competitiveness. The Indian government needs to encourage investment by organized sector to have large set up in weaving & processing. The effort should be on making good quality Indian fabric available at competitive price to apparel manufacturers.

The efforts are required to change the perception of buyers towards Indian fabric. The integrated effort of textile (fabric) and apparel industry shall boost all segments of textile and apparel industry. The government is also required to provide infrastructural support and policies, which can attract investment in this sector so as to develop all segments of industry. Competitiveness of Indian textile and apparel industry can be increased by strengthening elements of value chain in textile and apparel industry by developing competitive advantage of each of the segment and having its contribution for value addition in further stages of production. The strategic planning for targeting to segments offering higher realization due to product innovation and differentiation in higher value segments is required with building up comparative advantage of having strength in the backward value chain.

Bibliography

Aaker, David A. (1992): *Developing Business Strategies*. John Wiley & Sons New York

Aaker, David A. V. Kumar and George S. day (1999): *Marketing Research, (6th Edition)*, John Wiley & sons (Asia) Printed at Replika press (p) Ltd., Delhi

Abrahams, Alan. (1993): *Asia and World Textiles-The Prospects*. Apparel International (7/8): 52-55

Amersey, P.N. (1989): *Fabric Availability for Garments and its Problem*. AEPC Times 3-10 (11/12): 12-14

Anson Robin and Paul Simpson. (2001): *World Textile and Apparel Trade and Production Trends*. Textile Outlook International, September

Anson, Robin (1994): *Ongoing impact of the Uruguay round on Foreign Trade*, Textile Horizons, June

Apparel Export Promotion Council, 1995-2004: *Handbook of Export Statistics*, New Delhi.

Au K.F. and N.Y.Chan (1990): *The World Textile and clothing Trade: Globalization versus Regionalization*. The Hong Kong Polytechnic University

Bagchi, Sanjoy. (1991): *Integration of textile trade into GATT: Developing countries perspective*. Pakistan Textiles Journal XL (8): 15-20

Balaram Anuradha, Surendra S. Yadav, Rajat K. Baisya (2003): *Competitiveness of Indian Apparel Export Firms: An Analysis of select Delhi based Firms*. Global Business Review, 4:1: 57-76

Banik, N. and Saurabh Bandopadhyay (2000): *Cotton Textile Industry in India, in the Wake of MFA Phase-out. Working Paper no. 9*, Rajiv Gandhi Institute for Contemporary Studies, New Delhi.

Baughman, Laura M. & Kara M. Olson (1997): *Prospects for Exporting textiles and Clothing to the United States Over the Next Decade*. International Textiles and Clothing Bureau, Geneva, March.

Bhardwaj, Sunder. G., Rajan P. Varadarajan & John Fahy. (1993): *Sustainable Competitive Advantage in Service Industries. A conceptual model and Research Propositions*. Journal of Marketing 57 (10): 83-99

Bhatia, Satinder (1997): *Indian Garment Industry in the Post-MFA Period. Occasional Paper 7*, IIFT, New Delhi

Bhattacharya, B. (1991): *Export Marketing Strategies for Success.* Global Business Press, New Delhi

————————————. (1999): *Non-Tariff Measures on India's Exports, An Assessment. Occasional Paper No. 16,* IIFT, New Delhi.

Bhattacharya H.P., (1999): *MFA- Possible implications of dismantling for Indian Exports.* Asian Textile Journal, October: 49-52

————————————(1992): *Global Garment Industry and Trade.* The Indian Textile Journal CII (5): 38-42.

Bheda, Rajesh (2003): *Managing Productivity in the apparel industry* CBS Publishers, India

Cable, Vincent. (1985): *The Future of the MFA.* Paper presented at the CEPS conference co-sponsored by The Economist publication Ltd., December.

Chadha, R.S. Pohit, R.M. Stern, and A.V. Deardorff, (1999): *Phasing out the Multi-fibre Arrangement: Implications for India.* Paper presented at the Second Annual Conference on Global Economic Analysis. Copenhagen.

Chandra, P. (1999): *Competing through Capabilities: Strategies for Global Competitiveness of Indian Textile Industry.* Economic and Political Weekly, Feb 27.

Confederation of Indian Industry (1998): *India's Textile industry – Building Competitiveness to Survive and Thrive.*

Day, G.S. and Reibstein, D.J. (ed), (1997): *Wharton on Dynamic Competitive Strategy.* John Wiley and Sons.

Debroy, B. (1996): *Beyond the Uruguay Round – The Indian Perspective on GATT.* Sage Publications, New Delhi.

Dholakia, Nikhilesh. Rakesh Khurana, Labdhi Bhandari and Abhinandan, K. Jain, (1978): *Marketing Management: Cases and Concepts.* The Mcmilan Company of India Limited, Delhi

Dickerson, K.G. (1991): *Textiles and Apparel in the International Economy.* Macmillan, New York.

Dickerson, K.G. (1999): *Textiles and Apparel in the Global Economy.* 3rd edition. Prentice-Hall, Inc.

Erzan, R. G. G. and Paula Holmes (1989): *Effect of MFA on Developing Countries Trade.* Seminar Paper No. 449, Institute for International Economic Studies, Stockholm, September.

Export-import Bank of India (1995):*Indian Garment Exports-Implications of the MFA phase out.* Occasional Paper No. 34. Bombay: India Book House.

FICCI, (2000): *Implications of the Phase- out of MFA on the Indian Exports of Textile and Apparel*

GATT (1984): *Textiles and Clothing in the World Economy.* Geneva.

GATT (1987): Updating the 1984 GATT Secretariat Study: Textiles and clothing in the World Economy. Geneva.

Gokhle, C., Vijaya Katti (1995): *Globalising Indian Textiles – Threats and Opportunities.* Tecoya Publication.

Goswami, Onkar (1985): *Indian Textile Industry, 1970-1984 – An Analysis of Demand and Supply.* Economic and political Weekly, 21 September.

Goswami, Onkar (1990): *Sickness and Growth of India's Textile industry: Analysis and Policy Options.* Economic and Political Weekly, 3 and 10 November.

Government of India (1999): *New Textile Policy 2000-2004.* Ministry of Textiles, New Delhi'. 12th November.

Hamel, G. and C.K. Prahlad, (1994): *Competing for the Future.* Harvard Business School Press, Boston, USA.

Hamilton, C. B. ed., (1990): *The Uruguay Round Textiles Trade and the Developing Countries, Eliminating the Multi-Fibre Arrangements in the 1990.* The World Bank.

Hughes, Richard (1995): *The Uruguay round: New Approach for the Textile and Clothing sector,* International Trade Forum, No. 4.

Hughes, Richard, (2000): *Apparel Export in the New Millennium.* ITC Geneva, A Paper presented at Vigyan Bhavan Conference, New Delhi, India.

Indian Cotton Mills Federation (2000): *Annual report of ICMF.* Mumbai.

Indian Institute of Foreign Trade. (1978): *Ready-made Garment Industry in India: A study of Problems and Prospect.*

Iyer, Parameshwaran. (1993): *Indian Textile and Clothing Exports in the Global Context.* The Textile Industry and Trade Journal (7-8): 17-21

Jackson, H.K. and Frigon, N.L., (1996): *Achieving the Competitive Edge-A Practical guide to World Class Competition.* John Wiley & Sons, New York.

Jain S.K. (1988): *Export Performance and Export Marketing Strategies – A Study of Indian Textiles.* Common Wealth Publications. Delhi

Johnnsen Hano and Page G. Terry (1995): *International Dictionary of Management.* Kogan Page: 246.

Joshi Pradeep (2005): *Competitiveness of India's Textile Industry in WTO Era.* Textile Asia, July

——————— (2005): *Globalization in The Textile Industry* Asian Textile Journal, June

——————— (1999): *Garment Exports: Search of Excellence in the New Millennium* Clothesline and Fashion & Beyond, June

——————— (1997): *Marketing Trends In Motion* Textile Industry of India, August

Joshi Pradeep & S.M. Ishtiaque, Sudhir K.Jain (2005): *Opportunities for Indian Textiles and Apparel Industry in a Boundary Less World* Asian Textile Business, Sept

——————— (2005): *Competitiveness of Indian clothing Industry in WTO Era"* Asian Textile Journal, December

——————— (2005): *Competitiveness of Indian Textiles in WTO Era* Indian Textile Journal, June

——————— (2005): *The Road Ahead,* Clothesline, June

Joshi Pradeep & S.M. Ishtiaque, Sudhir K.Jain (2005): *The Preparedness of India's Textile and Clothing Industry for the Post Quota Era* Textile Asia, April

————————————(2005): *Indian Textiles and Clothing Industry: Prospects in Post Quota Era* The Textile Magazine, April

———————————— (2005): *Post WTO & the Indian Clothing Industry* Apparel Fortnightly (March 16-31)

———————————— (2005): *Global Textile and Clothing Trade Arrangements: A Historical Perspective of Transition to Boundary Less World* Fashion And Beyond, July

———————————— (2005): *Globalization in the Textile Industry: A Study of preparedness of Indian industry in WTO era* Paper at International Conference on "Advances In Textile Materials Technology, Management and Applications" at India

———————————— (2005): *Globalization of The Clothing Industry-A Study Of Prospects Of Indian Industry In WTO* Era Paper at International Conference on Apparel and Made-Ups Industry, India

———————————— (2005): *Globalization in the apparel Industry - A Study of preparedness of Indian industry Paper* at IIFTI conference 2005, Bunka University, Japan

Kapoor Hari. (1990): *EEC in the nineties-Greater Scope for India.* Clothesline 3 (8): 23-25

Kathuria S & Bhardwaj A. (1998): *Export Quotas and Policy Constraints in the in Textile and Garment Industries.* SASPR, World Bank, New Delhi Office, 27th May.

Kathuria, Sanjay et al. (2001): *Implications for South Asian countries of Abolishing the MFA,* revised version of paper presented at the NCAER-world bank, South Asia workshop, Dec. 20-21, 1999, New Delhi.

Katti V. (1996): *Textile Industry: Export Vision 2005.* Prepared for Expert Group on Textile Exports, Ministry of Textiles, New Delhi, August.

Katti, V. and Sen, S. (2000): *MFA Phasing-out and Indian Textile Industry.* Foreign Trade Review, October-March, xxxIV (3&4)

Kay J. (1996): *Business of Economics.* Oxford University Press, New York

Khanna, Sri Ram (1991): *International Trade in Textiles: MFA Quotas and a Developing Exporting Country.* ICRIER, Sage Publications, New Delhi.

———————————— (1993): *The New GATT Agreement* – Implications for the World Textile and clothing Industry. Textile Outlook International 52(3): 10-37.

————————————. (1992): *Profile of the Indian Textile and Clothing Industry.* Textile Outlook International 39 (1): 55-75

Khanna, Sri Ram and IBC Research Team, (2002): *Prospects for Fibre, Textile and Apparel Markets in India.* Jan-Feb, Textile Outlook International pp.62

Khurana, I.J.S. (1987): *Production Planning for Export of Ready Made Garments.* AEPC Times. (9): 6-9

Koschnick J. Wolfgang (1995): *Dictionary of Marketing.* Gower: 420-478

Koshy, Darlie O. (1995): *Effective Export marketing of Apparel to US, EU and Japan*. Global Business Press, Delhi.

———————— (1997): *Garment Exports: winning strategies*. Prentice Hall of India. Delhi

Kriplani, V.H. (1984): *International Marketing*. : Prentice- Hall of India. Delhi

Kurt Salmon Associates (1999): *Vertically Integrated Textile & Apparel Mission for India,* KSA Technopak, May.

McKinsey & Co. (1997): *Capturing Global Markets: A Strategy for Indian Companies*. Paper presented at Texcon Conference IIM Ahmedabad, India.

Mehta Rahul, (2001): *Exporting High Value Garments From Indian – Prospects*. Clothesline (October)

Ministry of Finance, Government of India (2000-05): *Economic Survey*

Ministry of Textiles, Government of India: Annual Reports (2000-2004). New Delhi

Misra, S. (1993): *India's Textile Sector- A Policy Analysis,* Sage Publications, New Delhi.

Modi, S.K., (2000): *India's Apparel Export Industry: Meeting the Challenge of a Quota-free Market*. Textile Outlook International, January.

Muchsin and Sudhir K. Jain (2002a): *Implications of WTO on Indian garment Exporters,* presented at national Seminar on Growth of Entrepreneurship in India during 19-20 2002, February, Jaipur.

Muchsin and Sudhir K. Jain (2002b): *Emerging Scenario of Garment Exports in the WTO Regime',* presented at 2002 NEBAA International Conference on Globalisation in the 21st Century, during May 30-31, Connecticut, USA, p. 16-18.

Neelamegham, S. (ed.) (1988): *Marketing in India: Cases and Readings* Vikas Publishing House Ltd. Delhi

NIFT-IBC (1999-2000): *Competitive Challenges Before the textile and clothing Industry'*. Report for Ministry of Textiles, Govt. of India.

NIFT- INDICA (2000): *Potential for Promoting Indian Apparel Brands in International Market '*. Report for Ministry of Textiles, Govt. of India.

Pankaj Chandra, (1997): *Competing through Capabilities: Strategies for Global Competitiveness of the Indian Textile Industry*. Texcon 97 – International Conference on Textile and clothing Proceedings, December.

Panthaki M.K. (2000): *Higher UVR holds key to garment export*. The Indian Textile Journal (August): 129-131

———————— (1999): *What Does the Future Hold?* Indian Garment Industry. *Clothesline*, March: 24-25.

Porter, M., Jeffrey D. Sachs & John W. McArthur (2001): *Executive Summary: Competitiveness and Stages of Economic Development*. Global Competitiveness Report 2001, World Economic Forum, Geneva

Porter, Michael E. & Ghemawat Pankaj (1995): *Developing Competitive Advantage in India*. CII, Harvard Business School, National Seminar, New Delhi, September 1994.

Porter, Michael E. (1980): *Competitive Strategy: Techniques for Analysing Industries and Competitors.* The Free Press. New York

———————— (1985): *Competitive Advantage: Creating & Sustaining Superior Performance.* New York, Free Press.

———————— (1990): *The Competitive Advantage of Nations.* The Macmillan Press Ltd, London and Basingstoke.

Prahalad, C. K., G. Hamel (1990): *The Core Competence of the Corporation.* Harvard Business Review, 168(3): 79-91.

Prasad, A.C.H. (1997): *India's Competitiveness in Export of Garments in the MFA Phase-out and Post-MFA Phase-out Periods,* Occasional Paper No. 10, Indian Institute of Foreign Trade, New Delhi.

Raffaelli M. (1998): *Bringing Textiles and Clothing into the Multilateral Trading System'.* in Bhagwati, Jagdish & Mathias Hirsch ed., The Uruguay Round and Beyond, Essays in Honour of Arthur Dunkel, Springer, Germany

Raffaelli, M. & Tripti Jenkins (1995): *The Drafting History of The Agreement on Textiles and Clothing.* International Textile and Clothing Bureau, Geneva, November.

Rajagopal, S. (1991): *Uruguay Round of Multilateral Negotiations implications for Textile Trade from India.* Conference Paper presented on GATT pages 110-118

Ramaswamy, K.V. and Gereffi, G., (1998): *India's Apparel Sector in the Global Economy – Catching up or Falling Behind.* Economic and Political Weekly, January, 17, Pages 122-130.

Roy P.R. (1997): *What is Holding Back Indian Textile Industry? – Industry Views, Strategic Options and Recommendations.* Texcon 97, Ahmedabad, December.

Roy Prodipto, (1999): *Competitiveness of the India Textile Supply Chain.* Textile Outlook International, September

Schiffman G. Leon and Kanuk Leslie Lazar (2000): *Consumer Behavior.* Prentice Hall of India, New Delhi: 146

Soloman R. Michael (1988): *Consumer Behavior.* Prentice Hall, New Jersy: 55-58

Stuart-Smith, Keith. (1993): *NAFTA, Uruguay Round, & Trade Agreements and their influence on World Trade in Textiles & Clothing.* Pakistan Textile Journal XLII (2) : 15-16

Sung, Kayser. (1993): *Asian Textiles 2000.* Pakistan Textile Journal XLII (8): 28-44.

Textile commissioner (2004): *Compendium of Textile Statistics.* Mumbai

———————— (2004): *Compendium of International statistics.* Mumbai

Textile Horizons (1994): The GATT Agreements: The Basic Aims, February.

Textile Outlook International (2000): *Trends in World Textile and Clothing Trade,* January.

———————— (2000): *Ready-made Garments in Export sector: High Cost Hurdle to Tech Upgradation,* July.

Textile Outlook International *(2000): Technology Upgradation Fund for RMG/ made-up units: Eligibility and List of Machinery*, September.

——————— *(2000): Bench marketing to combat Quota Phase-out Challenges*, December.

The Oxford Dictionary for International Business (1998): Oxford University Press

Trela, Irene and John Whalley (1990): Unraveling the Threads of the MFA CB Hamilton (ed), World Bank.

Varma, A.R. (1997): *Looking back at the Origin Value Addition in Textiles*, clotheslines, September.

Varma, M. L. (1995): *International Trade.* Vikas Publishing House Delhi

Varma, samar (2000): *Export Competitiveness of India Textile and garment industry, Working paper No. 9.*

——————— (2000): *Restructuring the Indian Textile Industry*, Vikas Publishing House, Delhi.

Yang, Y., W. Martin and K. Yanagishima. (1997): *Evaluating the Benefits of Abolishing the MFA in the Uruguay Round Package.* in T. Hertel (ed.), Global Trade Analysis: Modeling and Applications. Cambridge, Massachusetts: Cambridge University Press.

Yi Li, Yao Lei and Edward Newton, (2003): *Competitiveness of China's TC Industry.* Textile Asia (January): 41-45

Zeithami, Valarie. A. (1988): *Consumer Perceptions of Price. Quality and Value: A Means-end Model and Synthesis of Evidence.* Journal of Marketing 52 (7): 2-22

Websites

http://goidirectory@nic.net.in

http://www.wto.org

http://www.europa.eu.int

http://www.intracen.org

http://www.otexa.ita.doc.gov

http://www.cirfs.org

http://www.worldapparel.plus.com

http://www.worldbank.org

http://www.bharattextile.com

http://www.textilesintelligence.com

http://www.ciionline.org

http://www.icmfindia.com

http://www.ksa-technopak.com

http://www.ficci.com

http://www.arvindmills.com

Index